Cities and Cultures

Cities and Cultures is a critical account of the relations between contemporary cities and the cultures they produce and which in turn shape them. The book questions perceived ideas of what constitutes a city's culture through case studies in which different kinds of culture – the arts, cultural institutions and heritage, distinctive ways of life – are seen to be differently used in or affected by the development of particular cities. The book does not mask the complexity of this, but explains it in ways accessible to undergraduates.

The book begins with introductory chapters on the concepts of a city and a culture (the latter in the anthropological sense as well as denoting the arts), citing cases from modern literature. The book then moves from a critical account of cultural production in a metropolitan setting to the idea that a city, too, is produced through the characteristic ways of life of its inhabitants. The cultural industries are scrutinised for their relation to such cultures as well as to city marketing, and attention is given to the European Cities of Culture initiative, and to the hybridity of contemporary urban cultures in a period of globalisation and migration. In its penultimate chapter the book looks at incidental cultural forms and cultural means to identity formation; and in its final chapter examines the permeability of urban cultures and cultural forms. Sources are introduced, positions clarified and contrasted, and notes given for selective further reading.

Playing on the two meanings of culture, Miles takes a unique approach by relating arguments around these meanings to specific cases of urban development today. The book includes both critical comment on a range of literatures – being a truly interdisciplinary study – and the outcome of the author's field research into urban cultures.

Malcolm Miles is Reader in Cultural Theory at the University of Plymouth, UK, where he convenes the Critical Spaces Research Group and co-ordinates the doctoral research methods programme for the Faculty of Arts.

Routledge critical introductions to urbanism and the city

Edited by Malcolm Miles, University of Plymouth, UK
and John Rennie Short, University of Maryland, USA

International Advisory Board:
Franco Bianchini   Jane Rendell
Kim Dovey   Saskia Sassen
Stephen Graham   David Sibley
Tim Hall   Erik Swyngedouw
Phil Hubbard   Elizabeth Wilson
Peter Marcuse

The series is designed to allow undergraduate readers to make sense of, and find a critical way into, urbanism. It will:

- cover a broad range of themes
- introduce key ideas and sources
- allow the author to articulate her/his own position
- introduce complex arguments clearly and accessibly
- bridge disciplines, and theory and practice
- be affordable and well designed

The series covers social, political, economic, cultural and spatial concerns. It will appeal to students in architecture, cultural studies, geography, popular culture, sociology, urban studies, urban planning. It will be trans-disciplinary. Firmly situated in the present, it also introduces material from the cities of modernity and postmodernity.

Published:
Cities and Consumption – Mark Jayne
Cities and Cultures – Malcolm Miles

Forthcoming:
Cities and Cinema – Barbara Mennel
Cities and Nature – Lisa Benton-Short and John Rennie Short
Cities, Politics and Power – Simon Parker
Digital Cities – Chris Benner
Cities and Economies – Yeong-Hyun Kim and John Rennie Short
Urban Erotics – David Bell and John Binnie
Children, Youth and the City – Kathrin Hörshelmann and Lorraine van Blerk

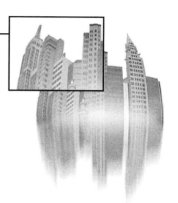

Cities and
Cultures

Malcolm Miles

Routledge
Taylor & Francis Group

LONDON AND NEW YORK

First published 2007
by Routledge
2 Park Square, Milton Park, Abingdon, Oxon OX14 4RN

Simultaneously published in the USA and Canada
by Routledge
270 Madison Ave, New York, NY 10016

Routledge is an imprint of the Taylor & Francis Group, an informa business

© 2007 Malcolm Miles

Typeset in Times New Roman and Futura
by Keystroke, High Street, Tettenhall, Wolverhampton
Printed and bound in Great Britain
by Antony Rowe Ltd, Chippenham, Wiltshire

British Library Cataloguing in Publication Data
A catalogue record for this book is available from the British Library

Library of Congress Cataloging in Publication Data
A catalog record for this book has been requested

ISBN10: 0–415–35442–0 (hbk)
ISBN10: 0–415–35443–9 (pbk)
ISBN10: 0–203–00109–5 (ebk)

ISBN13: 978–0–415–35442–4 (hbk)
ISBN13: 978–0–415–35443–1 (pbk)
ISBN13: 978–0–203–00109–7 (ebk)

Contents

Illustrations

Figures

Tables

Case studies

Acknowledgements

I am grateful for support from the Arts & Humanities Research Council (AHRC) which enabled me to visit sites and projects in researching this book. I am grateful also for the time and expertise of those with whom I have had conversations while writing the book. Among them are Vardan Azatyan, Sarah Bennett, Franco Bianchini, Iain Borden, Daniela Brasil, David Butler, Mario Caeiro, Sarah Carrington, Paul Chatterton, Tim Collins, Matthew Cornford, David Cross, Graeme Evans, Murray Fraser, Freee, John Goto, Reiko Goto, Paul Gough, Tim Hall, Angela Harutyunyan, William Hazell, Andy Hewitt, Sophie Hope, Mark Jayne, Mel Jordan, Nazareth Karoyan, Nicola Kirkham, Laima Kreivyte, Katy MacLeod, James Marriott, Steven Miles, Lucy Milton, Mindaugas Navakas, Barbara Penner, Jane Rendell, Marion Roberts, Judith Rugg, Esther Salamon, Gregory Scholette and Jane Trowell. I would also like to thank my series co-editor John Rennie Short and Commissioning Editor Andrew Mould and his assistants for their help and encouragement.

General introduction

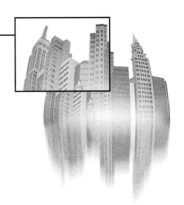

Rationale and scope

The terms *city* and *culture* indicate complex conceptual fields. Each figures in several academic disciplines, has its own history as a concept apart from the histories of specific cities and cultures and is used in various and at times conflicting ways. One of my aims in this book is to clarify how different uses of such terms indicate different assumptions. A second is to introduce readers in a range of academic fields to literature from fields other than their own. A third is to complement existing sources by adding a volume which is critical and introductory, and accessible without masking complexities. The book is intended for undergraduates in years 2 and 3 in urban studies, architecture, art and design, heritage, cultural studies, cultural policy, urban planning and sociology as well as cultural, human and urban geography; and postgraduates in interdisciplinary fields. I hope its juxtaposition of diverse sources, cases and ideas will enable new insights among readers, encourage reflexivity and contribute to discussion of urban conditions and future prospects.

The book plays on several ambivalences, and I address these in the first two chapters, on cities and cultures – which together act as a foundation for the rest of the book. Part of my aim is to draw out the mutability of such terms. The term *culture*, for instance, can mean the arts or a way of life. In culturally led re-development the arts are used to rebrand the built environment of city after city in the affluent world, but with mixed impacts on the broader cultures of dwellers. This is discussed in general terms, with the concept of the culture industries, in Chapter 5; and in a specifically European context in Chapter 6. Before that, in Chapters 3 and 4, I consider cities as sites of innovative cultural production, and how a city's image trades on selective cultural histories. In the last section of the book I move to questions of transition in the cities of the former East bloc after the end of the Soviet Union, of urban cultural identities, and of the permeability of built, social, economic, political and cultural environments.

Relations between the form of a city, its image in the arts and media, and the lives of dwellers may be conflictual. I show the complexities of this through contrasting cases, and include cases from a wide geographical spectrum as well as from different periods. Although there is a bias towards recent Western cases within the main texts of Chapters 2, 3, 4 and 6, Chapter 1 refers to one of the earliest cities in archaic Anatolia while other chapters look to the Americas and Eastern Europe as well as a peripheral, multi-ethnic zone in a Western European city. In case studies within each chapter (in a structure explained below) I balance eight examples from European cities (four dealing with popular culture) with five from the Americas, seven from Asia and the Near East, three from Africa, one from Australia and three relating to non-specific, literary sites. In case studies citing the work of other writers I have achieved an approximate gender balance, with significant representation of non-white commentators (within the book's anglophone scope). I use material from the media, fiction and personal accounts as well as from more academic sources in the arts, humanities and social sciences because each kind of source offers a specific kind of insight into a place or set of circumstances, and to tacitly raise the question of what is imparted in a particular (first- or third-person) voice.

The range of cases and theoretical material in the book reflects a diversity of cities in contemporary experience. I have tried to avoid a privileging of urban form in favour of recognition that the event of a city, so to speak, is produced in the continuing and always incomplete occupation of its built environment for dwelling, working, playing, buying and selling, social activity and political life. A city is, then, a location of daily life influenced by economic, political, social and cultural conditions. I provide no ultimate definitions but argue that the relation between cities and cultures is dialectical. That is, the conditions constituted by built, social, economic, political and cultural environments are influential on how people act in but also react to them, so that their reactions may re-order the conditions by which they are conditioned.

Structure

The book is arranged in four sections. Part One (Chapters 1 and 2) deals with the concepts of city and culture, playing on the dualities of meaning produced by their singular and plural forms while referencing a breadth of sources and intellectual frameworks relevant to the book as a whole. I link an account of the archaeological evidence of Çatal Hüyük, one of the first cities, to current discussion of the primacy of cities as human settlements. In Chapter 2 I outline a history of the term *culture*, and anchor discussion through the case of Tate Modern as a major cultural institution. Case studies range from an account of the

Spanish conquest of Mexico to a case for ordinary cities and a commentary on a photograph of pigeon lofts in north-east England.

Part Two (Chapters 3 and 4) reconsiders the role of cities as nodes of cultural innovation. Most cultural production and reception takes place in cities, and I relate this to the condition of proximity discussed in Chapter 1. I take the example of Paris because this is where key currents in modernism occurred, but ask if this is a relation specific to modernism and whether new patterns now emerge. Chapter 4 complements this by asking how cities are re-presented culturally, taking the case of Barcelona as a city which has transformed its image through cultural means. Case studies include accounts of allotment culture, San Francisco in the 1960s and the impact of gambling on Melbourne's recent waterfront development.

Part Three (Chapters 5 and 6) concerns the culture (or cultural and creative) industries. I compare the positions of contributors to debate over the past two decades in Europe and North America as well as with the arguments of critical theory from the 1930s to 1960s, and look at Disney's venture into urban design and management at Celebration, Florida. In Chapter 6 I move to the specifics of the European Union's programme of Cities (or Capitals) of Culture, citing Glasgow and Bergen. In case studies I cite writing on amateur film clubs in Poland, theatre in Eritrea's struggle for national liberation and the use of fiction for cultural aspirations in Singapore.

Part Four (Chapters 7 to 9) investigates everyday creativity and identity formation in local and global contexts, and the impacts of collisions between cultural frameworks. In Chapter 7 I focus on the cultural ferment of eastern Europe after the cultural, economic and political mechanisms of the Soviet era were abandoned. The survival of cultural signifiers, such as the Soviet-period statues transported to a forest park in Lithuania, raises questions as to how readings of a regime's public monuments shift when its power is over. Chapter 8 moves to a project in a multi-ethnic social housing neighbourhood of Lisbon, to ask how people use cultural means to construct identities outside mainstream cultural consumption. In Chapter 9 I emphasise the permeability of cultural forms through the derivation of the urban grid, looking at its development in Latin America and Spain. In case studies I cite cases from India, Armenia and North America as well as a reworking of the Robinson Crusoe story as a critique of received ideas of culture and building as vehicles of civilisation.

Format

Each chapter has the same format: a key question is addressed through theories and histories; a main case based on an urban site is discussed through the work of several authors and sometimes my own research. Three case studies offer 500-word commentaries neither to add to the arguments advanced nor to question them. The first is from journalism, fiction or a personal account. The second is an extract from an academic text. The third is a photograph, in most cases one of mine but in two chapters provided by artists. The oblique or contesting relation of case studies to the main texts is a tactic to encourage readers to reassess their own ideas and enliven seminar discussions based on the book. Each chapter begins with a set of learning objectives, and ends with a brief summary of its main points. I do not predict the learning outcomes of readers, seeing such an exercise as an unnecessary closure. Notes for further reading are collated at the end of each section, many of the titles cited being relevant to the section as a whole. A bibliography appears at the end of the book, with an index.

Approach

I write in the first person and use quotations to bring in the voices of other writers. This allows me to have a position within the field while acquainting readers with positions other than my own. Because I keep the text concise some compression is unavoidable, and at times reference will be necessary to the further reading suggested. I avoid notes because it seems better in an introductory work to decide whether to include material or leave it out, rather than add the qualifying remarks and asides I would use in a longer or more specialised book.

The book is an introduction but I have tried to avoid a reductive account. The term *city*, for instance, universalises aspects of cities as they are or might have been, often to normalise certain readings of the present. Since I have no belief in objective knowledge and want to avoid such normalisations, I use the plural of a concept to offer an open-ended reading of it. By asking what is distinctive about city lives rather than city forms, too, I emphasise habitation over design, on the grounds that meanings are negotiated in a city as well as streets, prices and living arrangements. Similarly, perceptions are moulded by values which dwellers as well as designers, authorities and commercial interests articulate. A site represents the outcome of a specific negotiation of space, technology, power and capital but there are always incomplete renegotiations in social, economic, political and cultural urban interactions. These are a few of the complexities within which I have attempted to produce a clear and concise but also critical introduction to cities and cultures. I hope that it will be informative and sometimes provocative, that it will encourage further reading (not only from the sources I suggest) and that reading it will be engaging.

Part One

Definitions

1 Cities

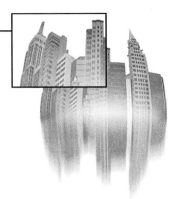

Learning objectives

- To understand what is distinctive about living in a city
- To outline some theories and histories which draw out particular qualities in urban living
- To refine the above by looking at a specific case, in which the quality of proximity appears a defining condition

Introduction

In this chapter I ask what is distinctive about living in cities. My concern is for qualitative rather than quantitative issues, and I dispense quickly with issues of data such as a city's size or the density of its population. But density leads me to the idea of proximity as the condition of a large mass of people from different backgrounds inhabiting a single site, which I develop through Edward Soja's account from archaeological evidence of Çatal Hüyük, one of the first cities, in Sumeria. Soja argues that at Çatal Hüyük the city was a primary form of settlement established before villages in its environs. Before looking at this I note the differentiation of urban from rural patterns of sociation in the work of early sociologists Emile Durkheim and Fernand Tönnies, and in Georg Simmel's view of metropolitan life; and contrast the work of Ernest W. Burgess and Louis Wirth in Chicago in the 1920s and 1930s with that of feminist writers such as Elizabeth Wilson (1991). Finally I ask if there is any substance in the idea that city air is liberating, looking to the writing of Henri Lefebvre. In case studies I cite Iain Sinclair's story of London's orbital motorway (2003); Jennifer Robinson's case for ordinary cities (2006); and an image of Hassan Fathy's experimental settlement at New Bariz, Egypt (constructed 1967, photographed 2005).

Concepts

Urban distinctions

Cities are large, densely populated human settlements, often with a degree of governmental autonomy from the national state. They have been hubs of trade and transport, defensive outposts and centres of manufacturing and cultural production. In the nineteenth century cities such as London, Paris, Madrid and Berlin were centres of political power reflecting the rise of nation-states in Europe, while Delhi, Algiers and Buenos Aires among others were nodes of colonial power and contestation – all centres of commercial expansion – and in the United States and Australia the function of political power was separated from that of commerce by the building of national capitals in Washington and Canberra. In the nineteenth and twentieth centuries, new movements in art, music and literature emerged in metropolitan cities such as Paris, Berlin, Munich, Vienna, Barcelona and St Petersburg when artists, writers and composers were attracted to milieux which also included critics and patrons. There are also university cities and cathedral cities, or cities which arose as major crossing points of a river or as ports, ancient cities such as Troy and Knossos known from their archaeological sites, and cities outside the dominant classifications of Western imperialism such as Djenne or Zimbabwe which were nonetheless nodes of trade and power. The difficulty is that as I adopt a range of variables I am presented with no more than a cumulative definition, while to fall back on legal and administrative arrangements or the granting of charters in medieval times is no more informative of what makes a city feel like a city today. I am drawn to ask instead if it is the availability of multiple dimensions of experience – the city as event, or cosmopolitan street life – that draws people to live in a city. But it is agricultural decline, rural poverty and the prospect of work that drives people to cities such as Lagos, Mexico City and São Paulo in less exalted ways today, as it did to London, Birmingham and Manchester in the English Industrial Revolution. Yet despite high costs people continue to live in a city such as New York because, I assume, they like it. Big cities are sites of anonymity in a crowd – a kind of freedom.

But what makes a city? Conventionally, villages have populations counted in hundreds, towns in thousands, cities in hundreds of thousands and metropolitan cities in millions (Angotti, 1993: 6). Today mega-cities have populations above twenty million (Soja, 2000: 236). But Thomas Angotti points out that a dense agricultural area can have a population equivalent to that of a small city (1993: 6); and Joel Garreau (1991) uses the term *edge city* for a spatially extended settlement with a large population but no centre which resembles a traditional city in name only. As planning increasingly looks to city-regions, and the idea of a city is so variable, size says little of what makes city life attractive. The factor of density is only slightly less equivocal. A received image of the European city is a knot of streets in which people live closely, a romanticised notion which ignores social distinctions and the hierarchies of streets. The term *city* has resonance, still, in contrast to suburb (not least in the marketing material of inner-city property developers). Perhaps the idea of a city links to that of a settlement protected against wild nature or invading armies, a safe place; or its resonance may derive from romanticised pictures of a medieval city with its towers, or the religious image of a city on a hill. The words *city* and *urban* both derive from Latin: *city* from *civis* (citizen), and *urban* from *urbs* (city). One implies an act of dwelling, the other a site. In French *cité* means a walled site and *bourg* an unwalled but protected area of land; both are separated from *commune*, a peripheric settlement for the poor (Sennett, 1995: 189–190) at least as densely populated as a *cité* but lacking its privilege and mystique. At the simplest level, a city is where things happen to influence history. For the privileged, a city reflects and allows expression of their status in a degree of governmental autonomy, as site of freedom from ties to the land, or site of commercial expansion and intellectual development. For the poor the city may be a site of oppression, as the land was before it.

Freedom was nonetheless material to the early modern idea of a city, housing trade which crossed the boundaries of principalities, free from the bonds of feudalism. It is not difficult to see how in eighteenth-century Europe a new commercial life and philosophy of reason were articulated in new city forms such as the grid plan (though I say more of this in Chapter 9). Cities were subject to rulers who had traditionally granted charters, collected tolls on movement and tariffs on goods and relied on a class of literate officials to administer their powers as well as artists and composers to establish the profiles of their courts. By the eighteenth century rulers had the power and technical resources to plan cities and then build them according to a plan. Yet the inscription over the gates to cities of the medieval Hanseatic League – city air makes people free (*Stadt Luft macht frei*) (Sennett, 1995: 155; Soja, 2000: 248) – retained currency for the bourgeois class. The Baixa district of Lisbon, the commercial centre rebuilt after 1755, with its marble pavements in designs of black and white and its arcades, expresses this commercially based freedom which challenged the power of religious and land-based elites. As Kenneth Maxwell writes, the new city was socially progressive:

> Pombal not only gave attention to the central squares and principal streets;
> more modest houses were designed and built as well, creating one of the first
> industrial development zones in a European city. Where the great aqueduct

terminated, Pombal placed his industrial suburb with silk manufactures, ceramic works, and cotton mills.

(Maxwell, 2002: 37)

The extension of Barcelona outside its walls is another such case, which I take up in Chapter 4.

Urban sociation

Commerce requires free contracts between individuals. In early sociology the kinds of association people make in cities are perceived as voluntary, distinct from the restrictive patterns available in village life. For Fernand Tönnies, the move to cities ends the obligations of a rural society when ties to kin and land are replaced by contractual associations between migrants (Tönnies, 1895; 1940; 1955). Emile Durkheim (1947 [1893]) sees the division of labour as key to the development of urban sociation, resting on a moral consensus enabling security of contracts between individuals of different status. This can be compared to Richard Sennett's account of the coffee houses of eighteenth-century Paris and London (1992 [1974]: 80–87; see also Rendell, 2002) where a specific mode of telling news bridged class backgrounds so that information was gathered as widely as possible. Commenting on Durkheim's *The Division of Labour in Society* (1947 [1893]), Peter Marris remarks that while his 'social prescriptions now seem quaint and dated' Durkheim's insistence on moral consensus is 'at the heart of all social democratic movements: the belief that economic relationships . . . must be governed by principles of justice and humanity' (Marris, 1998: 15). Tönnies and Durkheim are worth rereading, but I suggest that their division of urban from rural reflects a temporal rather than a spatial trajectory – a development from feudal to modern society rather than a comparison of different experiences within modern societies in which the countryside tends to be a source of food and recreation for city dwellers, a historical shift based on an assumption that cities offer more advanced social development than villages. The progressive trajectory echoes a utopianism in late nineteenth-century planning which follows a pronounced gendering – city as advancing masculine, countryside as retreating feminine – to become a patriarchal urbanism.

Utopianism, however, is a key factor in modern city development, in different ways as the following examples indicate. Ebenezer Howard's garden city is a synthesis of rural health and urban opportunity (Hall and Ward, 1998), and retains a central hub of cultural activity. Its philanthropic model is that of model villages for working-class housing at Bourneville and Port Sunlight, and translates with difficulty into fully urban terms. Camillo Sitte's plan for Munich, in contrast, sets out new city districts which reproduce the city as a whole with multi-use zoning, so that a city becomes a network of what might now be called urban villages (Meller, 2001: 43–44). In Idelfons Cerdá's plan for the extension of Barcelona (Soria y Puig, 1999) all classes are provided with decent housing, transport and green spaces, and half the space of a street is pedestrianised. It is remarkably progressive today. This utopianism transmutes in the 1950s into the work of the International Congress of Modern Architecture (CIAM) and the efforts of some of its (mainly better established) members to design a better world for urban-industrial societies (Curtis, 2000). Underpinning CIAM's work is an idea of a city as the most advanced form of settlement, the most modern in a circular definition in which the most modern is the most urban and technologically driven. This privileges expertise, as in the rational comprehensive planning model adopted in North America in the postwar period (Sandercock, 1998: 87–91). But cities are also where new alliances of workers were made. The first trades unions arose in the agricultural not the industrial revolution but the strength of unions in the twentieth century echo the scope for organisation available in industrial workplaces in cities. The growth of mass political movements similarly follows the existence of a critical mass of people. But if the Bolshevik Revolution of 1917 began in factories (Smith, 2002: 29–30), this does not mean that cities are the only places where a new society could be forged. Charles Fourier proposed to restructure French society through *phalansteries* – communities of equal size, spaced equidistantly – in which artisan and intellectual were to co-operate to produce a society based on the libidinisation of work (Beecher and Bienvenu, 1983). And today the conditions of village life include access to global communication technologies, blurring the distinction between rural retreat and inner-city loft. Farha Ghannam notes that immigrants from rural Egypt retain their former social ties and culture after relocation to Cairo in a search for stability (Ghannam, 1997), but in the affluent world a picture of rural life as stable and urban life as disruptive is fanciful. R. E. Pahl sees the idea of village as site of community as 'nearer to the middle-class image of a village' than to reality (Pahl, 1975: 41), while the notion of a place-based community is questionable in a period of global migration (Albrow, 1997; Fennel, 1997; Albrow, 1996: 155–159). This seems not to inhibit developers keen to promote a concept of the urban village as site of a place-based community, usually in post-industrial cities.

The rhetoric of urban development which extols the urban village as a way to describe gated apartment blocks extends a logic of improvement beginning in the seventeenth century, which gained prominence in the nineteenth. Victorian reformers saw the provision of public health as part of a new moralism, just as modern planners saw the provision of decent public housing as social restructuring designed for stability. Citing Michel Foucault (1988: 67), James Donald argues that the modern state uses two rationalities in maintaining its power: that of reason of state, or governance; and that of policing, which classifies and identifies a population. Donald writes: 'The population became the target of both surveillance and welfare' (Donald, 1999: 32). I would take map-making, of which there was an expansion in Europe in the seventeenth century (Pickles, 2004), like the statistical work to which Donald refers, as an example of technologies of policing. Donald continues, citing Paul Rabinow (1989), that 'this statistical grid of investigation produced a new social scientific conception of society that in turn informed the physical structure of the city' (Donald, 1999: 32). These references form a context for Chicago urbanism, discussed below, but the idea of a city as zone of increasing control is countered in Georg Simmel's writing, by which the Chicago sociologists were also influenced.

Urban stimulations

For Simmel the city is a site of over-stimulation and anonymity. Metropolitan Berlin becomes a state of mind. David Frisby links Simmel's work to Tönnies's idea that social evolution is spontaneous disaggregation (*désagrégation spontanée*): 'the social theorist is presented with the distinctive problem of locating and capturing the fleeting and the transitory' (Frisby, 1985: 46). In his essay 'Metropoles and Mental Life' Simmel writes of an intensification of perceptions, a 'sharp discontinuity in the grasp of a single glance' (Simmel, 1997: 175):

> Precisely in this connection the sophisticated character of metropolitan psychic life becomes understandable – as over against small town life which rests more upon deeply felt and emotional relationships. These latter are rooted in the more unconscious layers of the psyche and grow most readily in the steady rhythm of uninterrupted habitations.
>
> (Simmel, 1997: 175)

This reiterates Tönnies's distinction between village and city; but Simmel constructs a metropolitan psyche which co-exists with a rural psyche but is more imbued with the abstractions of intellectual life. He links this to the money economy, characterised as individualist while breeding a withdrawal from close ties. He links introjection of money with a flatness of metropolitan experience amid over-stimulation of the senses (nerves, as he puts it) resulting in 'an incapacity . . . to react to new sensations with the appropriate energy' – a blasé attitude – a 'colourlessness and indifference' which 'hollows out the core of things, their individuality, their specific value, and their incomparability' (Simmel, 1997: 178). The individual is less guarded than in a small group, and Simmel identifies 'the individual's full freedom of movement in all social and intellectual relationships' (Simmel, 1997: 184) as a key dimension of metropolitan life as crowds liberate people from inherited identities. This may link in ways which may not be compatible to Iris Marion Young's affirmation of group difference (1990), Louis Wirth's view of the city as a competitive jungle, and the idea of *flâneurie* in Charles Baudelaire's writing (see Chapter 3). Simmel continues that metropolitan conditions produce a reserved attitude: 'we frequently do not even know by sight those who have been our neighbours for years' – while 'This reserve . . . grants to the individual a kind and an amount of personal freedom which has no analogy whatsoever under other conditions' (Simmel, 1997: 179–180). This restructures a tendency to form small and close-knit groups, but in place of family the new ties depend on engagement in a society which Simmel, like Durkheim, sees as based on the division of labour.

Urban transitions

The idea of a city as site of freedom is problematic and gendered but it gives a hint as to why cities are regarded as vibrant when undercurrents of freedom threaten the agencies of policing. But Louis Wirth sees incoherence in a city left to its own devices:

> Cities generally, and American [*sic*] cities in particular, comprise a motley of peoples and cultures, of highly differentiated modes of life between which there often is only the faintest communication, the greatest indifference and the broadest tolerance, occasionally bitter strife, but always the sharpest contrast.
>
> (Wirth, 2003: 102)

He echoes early sociology in noting 'the weakening of the bonds of kinship, and the declining social significance of the family' in cities and goes on to say that urban dwellers are drawn to spectacle as 'escape from drudgery' (Wirth, 2003: 103). He then writes:

> Being reduced to a stage of virtual impotence as an individual, the urbanite is bound to exert himself [*sic*] by joining with others of similar interest into organized groups to obtain his ends. This results in the enormous multi-plication of voluntary organizations directed toward as great a variety of objectives as there are human needs and interests. . . . While in a primitive and in a rural society it is generally possible to predict on the basis of a few known factors who will belong to what and who will associate with whom . . . in the city we can only project the general pattern of group formation and affiliation, and this pattern will display many incongruities and contradictions.
>
> (Wirth, 2003: 103)

Wirth sees group formation as instrumental in a city but in a negative way as unable to provide a vision of a holistic future. For me his work precurses 1990s dystopianism (Davis, 1990; Smith, 1996) in his references to 'mental breakdown,

Case study 1.1

From Ian Sinclair, *London Orbital*

The buildings along Western Avenue don't want to be there; they'd prefer Satellite City. Or Las Vegas. Phoenix, Arizona, with Scunthorpe weather. They'd like to be closer to Heathrow's lingua franca. Mediterranean green glass. Low level units with a certain lazy elegance. *Super-Cannes* functionalism interspersed with Fifties grot. The heritaged emblems of an old riverside pub, The Swan & Bottle, have been banished by their corporate operators, Chef & Brewer, to the top of a wooden pole. That stares insolently at the slick shoebox of: X (The Document Company XEROX). The Xerox building is designed to look like office machinery, a shredder or printer. The windows are an enigmatic blue-green. Like chlorine. Xerox, Western Avenue, is a swimming pool on its side; from which, by some miracle of gravity, water doesn't spill. That's the concept: intelligent water. X marks the spot. Uxbridge is made from Xs. Lines of cancelled typescript. Fields planted with barbed wire.

(Sinclair, 2003: 218)

Comment

Sinclair's London, viewed from walking the approximate route of the orbital motorway M25, has little to do with a received concept of the city. Everything is out of place, better somewhere else, broken, or residual, never grand. He plays with words: 'The Xerox building duplicates itself' and sees a metaphor for the river in drifts of 'aqueous green glass' (Sinclair, 2003: 219). The landscape is transitional, a zone of roads to an airport and factories which use its transport links. This kind of landscape is often a visitor's first experience of a city, overlooked except for commercial property development where land prices are lower than in the centre. But in these wastelands there are houses, and hence neighbours. This no-one's land is populated: 'Workers and drones, in thrall to the glass beehives, plod down Slough Road, towards the UXBRIDGE sign' (ibid.). This is the end of the Metropolitan Line but not a metropolitan place, with little of the over-stimulation of which Simmel wrote. Sinclair adds, 'The women carry a second bag, slung from the shoulder, for personal effects' (ibid.). There are not many clues in Sinclair's account to the city's function of liberation from family or land-based ties; nor its superiority to village life; and little more privacy among the male and female work groups than among equivalent groups of agricultural workers. Imagination turns only to fantasies of flight to Phoenix or Las Vegas which reflects, or interprets, a mundane dullness and sense of need for something else. Or it might be homely after all.

At the canal he remarks 'Boat people keep their heads down, mind their own business – which is often survival (the new subversion)' (Sinclair, 2003: 221). Maybe this constitutes a counter-dimension to that of architectural spectacle in glass and steel. Or to the non-places of corporate and transport architectures which reproduce their visual codes globally. I spent the first eleven years of my life in a terraced house under the flightpath to Heathrow airport, near the Western Avenue on which Sinclair writes in this extract. I loathed the more respectable suburb to which my family moved in 1961, and had to get away to find the city. When, as a student, I could choose where I lived I found a room near the centre of London; and in the first phase of my academic career I lived in Maida Vale, a district of apartments with hardly any houses which, with the pastry shops and a population who settled there from middle Europe in the 1930s, made it seem un-English as well as non-suburban. I have happy memories of playing on wasteground as a child near the airport, and walking to the perimeter fence to watch the planes take off. Often my collaborator digging tunnels for toy cars in banks of earth on a building site was a boy from two doors along the street who now teaches critical theory at the University of California. But I never did find the city and have recently moved to Totnes – a small town in Devon set amid green, wooded hills above a sparkling, swirling river on which the perpetual sunlight glints.

suicide, delinquency, crime, corruption, and disorder' (Wirth, 2003: 104) – which strengthens the case for experts as arbiters of urban futures when all else threatens chaos. Michael Peter Smith summarises Wirth as saying in effect that 'Self-government is reduced to whatever emerges as a result of the unstable equilibrium of contending pressure groups' (Smith, 1980: 20). Wirth writes abstractly that individuals 'detached from the organized bodies which integrate society' make up the 'fluid masses that make collective behaviour in the urban community so unpredictable' (Wirth, 2003: 101). His urban type – the mobile, self-interested urbanite who is indifferent to others' needs – is a prescriptive projection, and Smith cites Wirth's view that interest groups intersect to produce conflict: 'To know one another better is often to hate one another more violently' (Wirth, 1964 [1948]: 329, in Smith, 1980: 5).

I object that Wirth naturalises competitiveness as a biological fact and foil to planning. Ernest W. Burgess also uses a biological metaphor for the evolution of transitional zones after waves of inward migration, 'analogous to the anabolic and katabolic processes of metabolism in the body' (Burgess, 2003: 160). His concentric ring diagram looks like a biological model, and is more prescriptive than descriptive (of Chicago). Mike Savage and Alan Warde say that writers mention it only 'to show that the subject of interest to them could not be fitted easily into its dimensions' (Savage and Warde, 1993: 16), but its repetition normalises its model of a central business district.

Chicago sociology had an ethnographic aspect as well, in F. Thrasher's *The Gang* (1927) and studies of hobos and dancing girls among other urban types. These deal more in actualities than Burgess's diagram, offering a less reductive image of a city's environment. For Savage and Warde, Thrasher's fieldwork using participant observation is a precedent for recent urban anthropology (Hannerz, 1980). Looking at the British experience, Ben Highmore writes of Mass Observation in the 1930s as 'mundane and poetic', gathering over a thousand reports of ordinary life from direct observation of, for instance, drinking in the public houses of Bolton and people's thoughts about margarine (Highmore, 2002: 75). But neither prescriptive diagrams nor the details of an individual's drinking habits give me an idea as to what is distinctive about living in a city, though perhaps this merely throws attention back to the inadequacy of the question in the first place – wanting a generalised answer while refusing the generalised concepts which might provide it. Yet there may be commonalities amid multiple experiences, and perhaps an idea of freedom.

Richard Sennett notes a model of the city as 'a place in which people could move and breathe freely . . . of flowing arteries and veins' (Sennett, 1995: 256), from Pierre Charles L'Enfant's 1791 plan for Washington DC as drawn by Andrew Ellicott in 1792 (Sennett, 1995: 266). Centres of power and social life are co-present in this multi-centred city, and the Mall was designed with the function of a lung as a place of social mixing that connected the parts of the city in which different groups of people lived (Sennett, 1995: 270). From this a public sphere in which a society determines its values and organisation through dialogue might be aligned with public spaces as sites of social mixing likely to be specific to cities. But this, though it requires the critical mass only cities offer, depends on the extent to which mixing involves people of different classes, races and genders. In Washington this was not the case, some of its inhabitants being slaves and others disenfranchised because they were female or poor, much as was the case in classical-period Athens. Perhaps the idea of a city as site of democratic exchange rests more on an Enlightenment mythicisation of the city reflecting the emergence of a bourgeois class, then, than on the actualities of city lives as might be revealed in social research.

Perspectives

From a critical reading of Chicago urbanism and more recent contributions such as Sennett's (1990; 1995) I could say provisionally that cities are complex environments in which social formations occur and re-occur, power and property are contested, and traces of an invisible architecture of sociation overlie the visible environment which itself bears traces of multiple concepts of a city effected by past generations. This gives me fluidity, and a recognition that cities are produced rather than simply designed – though the symbolic idea of a city remains influential on aspects of urban development, for example in the tendency today for urban designers to provide new public spaces almost as badges of respectability for development schemes. It is not surprising that people from different class, gender and ethnic backgrounds have historical reason to regard the idea of a city and its relation to a possibly mythical freedom differently.

Peter Marcuse demonstrates that cities comprise overlapping zones of business, power, industry and housing within all of which are distinct zones, to produce a 'clustering along a number of dimensions, some related to and congruent with others, other lines independent of all others' (Marcuse, 2002: 102). This approach can be extended for analysis of urban cultures to draw attention away from city form towards the numerous acts of habitation which are its everyday actuality more or less regardless of any grander design. Henri Lefebvre argues in *The Production of Space* (1991) that the unified space of design is always interrupted by a dimension of constantly remade meanings and associations in the use of urban spaces – the lived spaces which are not bound by, even if they are imposed on by, the unified space of plans and elevations. I return to this idea below but at this point turn in particular to the gendering of the visible city of public spaces, public monuments and grandiose buildings. Here the force of cultural norms is apparent in a privileging of the visual over the tactile or aural, and with it arguably a privileging of certain social and political attitudes and assumptions.

Marsha Meskimmon (1997) argues that cities in modernity tend to be visually spectacular, and Doreen Massey (1994) that visuality is a way of perceiving a city which privileges an authoritative masculinity. Massey writes that 'The privileging of vision impoverishes us through deprivation of other forms of sensory perception' and that 'the reason for the privileging of vision is precisely its supposed detachment' (Massey, 1994: 232) – which is always from a specific viewpoint. Citing Griselda Pollock, who argues (1988) that it is inadequate to add women's art to men's art history when a new art history based on a new taxonomy is needed, Massey says that modernism reflects a masculine sexuality which objectifies women's bodies. The conventional city plan could be critiqued as a bird's- or god's-eye view of a city, an example of how visual culture neutralises what is on closer investigation a contentious approach. Meskimmon observes that modern artists depict a city's interstices, sites of transgressive frisson such as the street at night, the bar or café, or brothel, more than monumental spaces. She writes, 'These spaces all represent dangerous meetings between the social classes

Case study 1.2

From Jennifer Robinson, *Ordinary Cities*

Instead of seeing some cities as more advanced or dynamic than others, or assuming that some cities display the futures of others, or dividing cities into commensurable groupings through hierarchising categories, I have proposed the value of seeing all cities as ordinary, part of the same field of analysis. The consequence of this is to bring into view different aspects of cities than those which are highlighted in global and world city analyses.

First, ordinary cities can be understood as unique assemblages of wider processes – they are all distinctive, in a category of one. Of course there are differences among cities, but I have suggested that these are best thought of as distributed promiscuously across cities, rather than neatly allocated according to pregiven categories . . .

Second, . . . ordinary cities exist within a world of interactions and flows. However, in place of the global- and world-cities approaches that focus on a small range of economic and political activities . . . ordinary cities bring together a vast array of networks and circulations of varying spatial reach and assemble many different kinds of social, economic and political processes . . .

(Robinson, 2006: 108–109)

The driving force of the economy [of Johannesburg post-1994] . . . gave strong encouragement to politicians and officials eager to make economic growth a priority. However, the diversity of the ordinary city made this difficult to achieve without attention to the wider city context. The needs of the poorest residents were not only electorally important, national urban legislation mandated developmental local government and demanded strong attention to delivering basic needs. Thus the strategic vision for Johannesburg had to adopt a more nuanced perspective . . . The economy, then, was portrayed as the underlying priority . . . but any political support for economic growth had to . . . include attention to basic service delivery.

(Robinson, 2006: 138)

Comment

Jennifer Robinson brings cities outside the loops of the global city (Sassen, 1991) and world-city (Olds and Yeung, 2004; Gilbert, 1998; Douglass, 1998) into mainstream discussion. Lusaka, Kuala Lumpur and Johannesburg are ex-colonial cities but the category includes all cities regardless of assignment to other categories. She argues that views privileging economic mechanisms ignore factors which determine how economic aims are delivered; and that a city's future is produced in complex interactions of social and political factors, global webs of communication, and economic forces and policies, as the example of Johannesburg (which she discusses in some detail) illustrates. Robinson challenges a tendency to write only about cities seen as predominant – whether for success in cultural rebranding like Barcelona, or dystopian conditions like Los Angeles – and taken as models on which the future of other cities is based. Robinson reads a 'strongly cosmopolitan imagination of possible urban futures' (2006: 148) in the making of a vision for Johannesburg in 2030. This uses aspects of global urban development such as the city's high-rise towers, and the emblematic structure of the apartheid-era Hillbrow Tower, to buy into global competition for a place on the map of international trade and tourism, and 'to stake a claim for a place in global cultural imaginations of modernity' (2006: 147); yet this is also a strategy for the city's re-presentation post-apartheid as a hub in Africa and the world. Robinson cites a Johannesburg politician as saying the models for the 2030 vision include Porto Allegre in Brazil, Birmingham and cases from Canada and Australia. Robinson observes that the vision rests on diverse histories mediated by a need to provide basic services to the city's population, redirecting attention to local conditions. This does not make Johannesburg an ordinary city in contrast to cities in which such needs are ignored, but makes it a specific case of urban development, and all the other cases similarly ordinary in representing diverse ways of negotiating an axis between a world-view and a set of conditions.

There is an axis, too, between planning and deregulation. Privatisation of public utilities has occurred in cities in the new South Africa though in some localities dwellers have refused to pay the private water companies. Robinson does not mention that, but her book reminds me that the issues are not unique in any city while the approach to them is uniquely determined by local as well as global conditions. The idea of ordinary cities is helpful in its insistence on diversity as the only generalisable condition of city living, hence the only basis for imagining a city's future. I like Robinson's remark on the Chicago School's generalisation of a duality of here/now and there/then: 'These fantasies enhanced urbanists' sense of elation about what they liked to think of as the novelty of their experiences . . . a most inadequate foundation for a post-colonial urban studies' (2006: 7–8).

or, more significantly here, the sexes' (Meskimmon, 1997: 9). It could be added that modernist artists and art movements rejected traditional perspective's single viewpoint in favour of multiple points of reference for multiple realities, as in Cubism and Futurism (though these movements remained male-dominated).

In feminism and deconstruction, discussion of cities moves from coherence to recognition of difference. Cities favour a heterogeneity which Elizabeth Wilson sees as their vitality, writing that Sennett was right in *The Uses of Disorder* (1970) that 'the excitement of city life cannot be preserved if all conflict is eliminated . . . that life in the great city offers the potential for greater freedom and diversity than life in small communities', adding that 'This is particularly important for women' (Wilson, 1991: 156). For Wilson planning regimes are detrimental to women's freedom in the city because to male planners women signify an irrationality which rational city planning cannot admit: '[Women] have seemed to represent disorder . . . civilization has come . . . to mean an authoritarian control of the wayward spontaneity of all human desires and aspirations' (Wilson, 1991: 157). There are alternatives: Leonie Sandercock recounts several models from advocacy planning in the 1960s to radical planning in the 1990s which bring planners into a dialogic relation with dwellers (Sandercock, 1998: 107–125). Jane Rendell, in *The Pursuit of Pleasure* (2002), reads city spaces as gendered sites which nonetheless admit women to new freedoms, arguing that 'Consumption can be seen as important in the lives of women . . . The role of consumer may be seen as an empowering one, a source of self-identity and pleasure in the public realm' (Rendell, 2002: 17, citing Dowling, 1993; Wilson, 1992). Instead of Wirth's city of relentless competition, then, is there a city of empowerment? For Kian Tajbakhsh, a city 'is not so much a territory or a place as it is a promise, a potential, built on an ethics of respect for the hybrid spaces of identities' (Tajbakhsh, 2001: 183). The idea of diversity may render concepts of the city redundant, to be replaced by localised analysis of cities, or discussion of difference as a pervading quality irreducible to a set of characteristics, visual or tactile.

Or I might look to what French economist and writer Jeanne Hyvrard calls *enception*: 'a thought process that the thinker is part of as it happens' (Waelti-Walters, 1996: 130 – see Chapter 8) to enable me to think about cities as entities in an endlessly adaptable and open network which produces an always-unfinished idea as to what the entities themselves are. I end this section by citing black poet Audre Lorde's remark at the Second Sex Conference in New York in 1979:

> *For the master's tools will never dismantle the master's house*, They may allow us temporarily to beat him at his own game, but they will never enable us to bring about genuine change.
>
> (Lorde, 2000: 54)

Definition is one of those tools, linear history another. This does not mean that only micro-scale studies are helpful but that meta-concepts of the city carry a weight of received assumptions and privileges which inhibits understanding of what cities are and what city lives might become.

Trajectories

From ape to moon-walker

An image of a city derived from medieval or archaic precedents as protecting its inhabitants against wild nature and invading armies is attractive, yet it is too easy to imagine a city as a citadel. I wonder if there is less deception in the idea of a city as culmination of a history of human settlement. The difficulty is the model of a trajectory, when most trajectories privilege the final term. In a progress of human settlement on these lines the city is the superior and the village the inferior term. The same model was used in early ethnographic displays in which the artefacts of colonised peoples were displayed, decontextualised in glass cases, beside the tools of early European societies (Coombes 1991; 1994a; 1994b). This rested on an alignment of cultural, technological and social progress, and a notion of evolution not as the adaptation to environment without design for which Darwin argued but as survival of the strongest, or realisation of an essential human nature.

A primary city?

The case of Çatal Hüyük refutes this trajectory, which is why Edward Soja's account of it is important. Çatal Hüyük in Sumeria, modern-day Anatolia, was built in wood and mud-brick at least 8,000 years ago and housed a population of around 10,000 – very large for the time. Soja opens his account of Çatal Hüyük by citing V. Gordon Childe's theory of geohistorical evolution:

> The very use of the terms 'geohistory' and 'cityspace' immediately reflects the preferential foregrounding of a critical spatial perspective. Geohistory, for example, emphasizes the unprioritized inseparability of geography and history, their necessary and often problematic interwovenness.
>
> (Soja, 2000: 7)

Soja draws on the work of archaeologists James Mellaart, who dug at Çatal Hüyük in the 1960s, and Ian Hodder, who resumed work there in the 1990s. He cites C. K. Maisels's (1993a) *The Emergence of Civilization: From Hunting and Gathering to Agriculture, Cities, and the State in the Near East*, with its trajectory of human social evolution from hunter-gatherer to farmer and city-dweller, in which the city-state is a move from barbarism to civilisation. Soja sees a rich geographical imagination in Maisels's work, and a concern for the dynamic potential of proximity. Soja adopts the term *synekism* from Maisels's *synoecism*:

> Synekism is directly derived from *synoikismos*, literally the condition arising from dwelling together in one house, or *oikos*, and used by Aristotle in his *Politics* to describe the formation of the Athenian *polis* or city-state.
>
> (Soja, 2000: 12)

Soja defines synekism as 'a fundamental and continuous force in the entire sequence of human societal development' (Soja, 2000: 26), but argues against Maisels's periodisation. Synekism is a foundational quality of urban dwelling but not in a temporal sense; Maisels, he says, 'presumes a . . . path in which small farming villages grow ever larger until some threshold of synekism is surpassed and true cities and city-states "crystallise" in very particular locations' (Soja, 2000: 25).

Soja notes that Maisels differentiates Sumerian city-states as articulating an ideology of hegemony and dependence from matrilineal chiefdoms and patrilineal villages maintaining a myth of divine descent:

> In these distinctions, the role of the city versus the village . . . features prominently. Following Marx, Maisels argues that in the Asiatic or village-state mode cities are 'exceptional' and 'quite unproductive,' royal 'islands in a sea of peasant villages.' Only in the city-states of Sumeria does full-fledged urban society and increasingly stratified class-like relations of production take hold.
>
> (Soja, 2000: 25)

Soja then cites a passage from Maisels which succinctly states the urban quality of the city-state:

> By *urban*, I mean a population sufficiently numerous and nucleated that the social relations of production mutate to express the principle of synoecism itself . . . the emergent expression of which is the crystallisation of government. In turn, government manifests itself as the state through administration, based on writing, plus monumental building representing the professionalization of ideological, economic, and armed force. It is not coincidental, then, that the first form of the city, or for that matter the state, takes the form of the city-state.
>
> (Maisels, 1993b: 155, in Soja, 2000: 25)

Cities are sites of innovation and propinquity, but for Soja do not evolve from previous forms. He proposes 'putting cities first, that is, pushing back the origins of cities to a time *before* the Agricultural Revolution' (Soja, 2000: 26). It is not a case of a growth of agriculture and village as precondition for the city but of cities being the primary settlement out of which villages emerge when the surrounding belt of cultivation and herding moves far enough out to require overnight or longer stays. This allows a reconceptualisation of the city as location of the proximity of a mass of people in one place 'inverting the usual chain of causality' (Soja, 2000: 27).

Among the finds at Çatal Hüyük is a mural of the city as a dense built fabric without streets (Soja, 2000: fig. 1.6). People moved from building to building via the roofs, using ladders to traverse levels (Mellaart, 1967: figs 59, 60). Soja calls it pueblo-like. I see it as close to the mud-brick villages of the Dra Valley or *ksours* of the Dades Valley in Morocco (Dethier, 1982: 18–19; Goldfinger, 1993: 122–149). Dwellers in Çatal Hüyük consumed a variety of foods. Storage bins denote the keeping of surplus grain, and there is evidence of a wide range of artisan trades – baking, weaving, basket-making, carving, tool-making, mirror- and copper-work and painting, from a long list given by Mellaart (1967: 99, in Soja, 2000: 39). There may have been gender equality. Mellaart writes: 'the position of women was obviously an important one in an agricultural society with a fertility cult in which a goddess was the principal deity' (Mellaart, 1967: 225). Hodder (Soja, 2000: 48, n25) is doubtful about the goddess cult but found a female figurine in 2004 which might support it (http://catal.arch.cam.ac.uk/catal/Archive_rep04/ar04_01.html). This adds up to an 'Urban Neolithic' which was 'hitherto an impossible juxtaposition of terms' (Soja, 2000: 42).

The first dwellers in Çatal Hüyük were hunter-gatherers whose previous timber houses are echoed in its timber-frame construction. In the background of the mural is a volcano, Hasan Dag, source of obsidian. Soja writes that nature is thus incorporated in the city's territorial culture and symbolic zone, 'as a vital part of the local economy and society, signalling the beginning of the social production of a "second nature" intricately involved with the urbanization process' (Soja, 2000: 42). Soja's interpretation follows Jane Jacobs in *The Economy of Cities* (Jacobs, 1969; Soja, 2000: 42–46), which refuted Lewis Mumford's position in *The City in History* (1961) – which Soja describes as a 'romancing of the pre-urban agricultural village' (Soja, 2000: 43). Soja writes that Jacobs 'put cities first, attached the formulation of agricultural and herding villages to the first cities, and reconstructed her own version of the city in history' (Soja, 2000: 43). The archaeological evidence supports a model of an agricultural revolution driven by the city's expansion (not its cause) and Çatal Hüyük establishes the city as a primary form of settlement.

Case study 1.3

Figure 1.1 *New Bariz, Egypt, planned and designed by Hassan Fathy (1965–67), market and agricultural co-operative building (photo M. Miles)*

Comment

New Bariz in the Kharga Oasis in Egypt is a 200-acre site designed by Hassan Fathy to house 250 families, built in mud-brick between 1965 and 1967 (when construction stopped owing to the Israeli invasion of Sinai). It was part of President Nasr's attempt to green the Western desert, renamed the New Valley, to reduce pressure on housing in Cairo. It was to have been a hub for six satellite villages. Irrigation allows cultivation of the fertile soil, and today fields of barley and pockets of brilliant green onions interrupt the desert landscape. The only buildings on the site now are a market complex for an agricultural co-operative (*suq*), three houses, a bus waiting room, unfinished administrative offices and a workshop for car repairs. There is no rain. New Bariz is well preserved.

Fathy studied the existing building types of the Kharga region, and drew on his work for Doxiades on housing projects in Iraq, and the discussions of the City of the Future project (Steele, 1997: 111–123). Since the inhabitants were unidentified, but he supposed would move to New Bariz when housing became available, Fathy was unable to carry out the social research he had attempted, unsuccessfully, at his better-known site at New Gourna (Fathy, 1969). James Steele describes the *suq* as 'the ultimate test of Fathy's attempt to ameliorate the extremely harsh conditions without mechanical means' (Steele, 1997: 141). Fathy used an orthogonal plan to achieve maximum shade in temperatures reaching 45°C in summer, with a planned (unbuilt) mosque angled obliquely to a palm-lined square. Natural ventilation, convection and the juxtaposition of open and closed spaces cool the interiors. The *suq* uses a row of wind-catches to draw air through the walls of the north side, to keep the temperature of a series of underground vaults for storage of vegetables for sale by the co-operative, at 15–18°C in summer. The building has a central courtyard, spaces for shops and a vaulted arcade. Fathy writes, 'the only alternative is to resort to the traditional cooperative system by finding means to make it work under the non-traditional conditions prevailing . . . one man cannot build a house, but ten men can build ten houses easily' ('Bariz Case Study', in Richards, Serageldin and Rastorfer, 1985: 92). Fathy recognised the absence of the building skills used at New Gourna, seeing a role for the architect as intermediary: 'it is implicit that we secure the assistance of the specialized architects to revive the lost experience and traditions among the peasants until a new tradition is established' (ibid.). He adds that the families involved should know each other, the optimum number of families in a collective being twenty. The question which lingers in my mind (see Miles, 2006) is whether Fathy's modernism, which I take as an alternative modernism in a nationalist context, still relies on the agency of professional experts or hands over the building and imagination of an urban future to dwellers. I leave that open here.

Spaces for living

The production of spaces

As Soja articulates in his use of the term *synekism*, human settlement houses collaboration. From a different viewpoint, more concerned with political forms, Amitai Etzioni writes of civil society as making a basis for social order in a civil culture of communitarian values translated into the mechanisms of civil ordering (Etzioni, 2004: 28). Although this retains a faith in state structures of control it qualifies that in terms of an ethical consensus which can underpin a self-organising society. Given the concern with proximity above I ask now to what extent the conditions of a city (such as proximity, enabled by high-density occupation) are conducive to the formation of such a consensus, and its articulation in the determination of social forms.

Some histories offer a negative perception of high density, seen more as a seed bed of unrest and deviance than of order. High-density housing was regarded in nineteenth-century London as fostering dirt, disease, crime and revolt, to be purged by architectural re-ordering as in the clearances which made way for Regent Street and Trafalgar Square in London, or the new boulevards of Paris. This follows previous efforts to purge city air of the miasma of the dead and odours of the living (Corbin, 1996; Illich, 1986). In the postwar period, partly but not only as a result of wartime destruction, high-density inner-city housing in major cities in Britain was replaced by suburban estates. Similarly in North American cities, high-rise towers of project housing set in green margins became the dominant form of social housing. Planners and designers on both sides of the Atlantic were committed to social benefit. Edward Robbins describes a planner's view of such areas: 'Social life could appear to the uninitiated to be endlessly enervating and debilitating; so much so that those living in such communities would find it difficult to improve their lot because of the surrounding environment. Poverty, idleness, and conflict would all appear to be in the nature of the place' (Robbins, 1996: 285). Robbins argues that inner-city streets were coherent, multi-zoned spaces for users who were told by implication in the new functionalised environments provided for them under the guise of decent housing (which they were in terms of sanitary conditions) that they had no ability to order space for themselves. This, rather than budgets or technologies, was the flaw in social housing schemes from the 1950s to the 1970s. An alternative, clearly, would be to recognise the spatial practices of dwellers as evidence of a degree of expertise more tacit than that of planners or designers but no less valuable in its own terms.

Lived spaces

To take this further I look now to the French Marxist thinker and urbanist Henri Lefebvre. Since *The Production of Space* was translated into English (1991), his influence on urbanism in the anglophone world has increased rapidly. His theory was first published in French in 1974, after many previous works on urban life and spaces, and consolidates his thinking in a fugue-like narrative in which themes and variations are stated and restated. To give one of the core findings he articulates, Lefebvre regards spatial practices – the ways in which people handle and negotiate spaces – as characteristic of a society at a specific time. Hence Roman spatiality is concerned with intersections and standardisation, a combination of straight lines and curves, and organisation of volume on functional lines: 'It embodies the simple, regulated and methodical principle of coherent stability, a principle operating under the banner of political religion and applying equally to mental and to social life' (Lefebvre, 1991: 238); and 'The resulting rationality, whether spatial or juridical, is detectable everywhere in the essential and most concrete creations of the Roman mind: vault, arch, circle . . . a spatial practice, mobilized by (equally spatial) representations' (Lefebvre, 1991: 244). Perceptions of space produce characteristic forms – one of the most immediate reference points in looking at the remains of past societies – and take place in two dimensions (by which I mean realms of understanding like an aesthetic dimension, and not geometric dimensions). One is conceived space, which Lefebvre terms the representation of space (singular); the other consists in representational spaces (plural), or lived spaces. The two overlie each other, and Lefebvre's tripartite theory is not a means to champion one over the other but to draw out, for the purposes of a dialectic, the qualities of each. Conceived space is the space of plans and elevations, geometries and quantitative ordering. Lived space is the realm of occupation in which people produce meanings and desires, call on memories and associations and re-order spaces. Again, this is not as I understand it a polarisation of rational and irrational spaces, since both conceived and lived spaces are inflected intuitively, but a recognition that the design of space according to a continuous field (in the abstraction of Cartesian geometry) is re-ordered by its occupation. This is practised characteristically, so that Roman representational spaces split a masculine dominance from a feminine 'thrust down into the "abyss" of the earth, as the place where seeds are sown and the dead are laid' (Lefebvre, 1991: 245). Using the term *intuitus* as the characteristic intuition of space, Lefebvre sums up:

> The triad perceived-conceived-lived . . . contributes to the production of space through interactions which metamorphose the original *intuitus* into a quasi-system: the vault and its magic, the arch, or the aqueduct. In the case of Rome, organization, thought and the production of space went together . . . under the sign of the Law.
>
> (Lefebvre, 1991: 246)

For ten years I have found Lefebvre's ideas illuminating, and refer to other aspects of his work in Chapter 9. Here I want to draw out the importance of lived spaces, not against conceived space but as antidote to the emphasis on conceived space in conventional planning, for the Chicago School based on a perception of urban dysfunctionality. Marshall Berman (1983) describes the impact of such a regime on the lived-in city under planner Robert Moses in postwar New York; and Lefebvre comments that to privilege abstract space via bureaucracy leads to 'an authoritarian and brutal spatial practice' (Lefebvre, 1991: 308), citing Baron Haussmann's remodelling of Paris in the 1850s and 1860s (see Chapter 3) and Le Corbusier.

It can be argued that the authoritarian spatial regimes against which Berman wrote are a product of large cities in which planning regimes deal with diverse publics they see as unable to form a value consensus; and that one of the new specialisms produced in conditions of proximity within a wider division of labour is urban planning itself. It can equally well be argued that there is evidence to the contrary, in everyday uses and adaptations of urban spaces by dwellers – from DIY to self-built but also from using a street in various ways to navigating the city by personal points of reference. I suggest that if lived space is the dimension of dwellers it is a realm produced in particular ways in cities, especially those large enough to house multiple publics.

Summary

I have argued that the account of Çatal Hüyük is significant in breaking the trajectory of farm village town city of nineteenth-century sociology, and from that the model of a linear, temporal development in general. The relation of proximity to division of labour is significant as well, enabling cultural produc-tion and political life to develop, and parallel to this abilities to adapt spaces in personal and shared ways according to personal and shared values from which arises a possibility to imagine or inflect urban futures. If I have a concluding point to offer it is that proximity offers new freedoms, and is a distinctive aspect of city living requiring the critical mass which produces diversity. If autonomy is more likely to arise in cities than elsewhere, the next question is whether this is a modernist aspiration or applies today in de-industrialised cities, edge cities or suburbs. But complementing that question would be acceptance that new technologies of information and communication make the geographical dimension of a city less important than the links of common interest of citizens (however effected, including via the internet as well as in a café or bar, or on a street corner), and the symbolic dimension of a city's image (discussed in Chapters 5 and 6).

2 Cultures

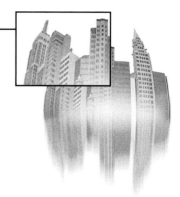

Introduction

In Chapter 1 I asked what is distinctive about living in a city. I saw the condition of proximity as enabling a division of labour, a sense of autonomy and new social relations among individuals (including the anonymity available in a large city). In this chapter I ask what the word *culture* means, citing Catherine Belsey's (2002) definition as a vocabulary within which people act and Raymond Williams's idea in *The Long Revolution* (1965) that social change occurs incrementally through cultural and educational changes. I draw on his definitions of culture in *Keywords* (1976), then look at Tate Modern as a major cultural institution. In case studies I cite a history of the colonisation of Mexico by Octavio Paz; an account of Samuel Mockbee's Rural Studio for grass-roots architecture in the southern USA; and an image of pigeon lofts in Skinningrove in the north-east of England. All these counter the potential dominance of Tate by emphasising the cultures of dominated peoples and overlooked places, implying that culture does not need taking to supposedly underprivileged publics because they already have it in plenty. This oversimplification, however, rests on a confusion of culture as the arts and culture as a way of life. I begin by setting out contested meaning for culture so that such confusions can be avoided.

Culture and cultures

Contested meanings

Catherine Belsey offers the following definition of culture in *Poststructuralism: A Very Short Introduction* (2002): '**culture**: the inscription in stories, rituals, customs, objects, and practices of the meanings in circulation at a specific time and place' (Belsey, 2002: 113). This definition operates at several levels, including that of cultural production (the arts) and the institutional structures which validate it, and the everyday acts of individuals and groups within a society (cultures). Belsey makes no distinction here between high culture and the cultures of every-day life. Earlier she writes:

> Culture constitutes the vocabulary within which we do what we do; it specifies the meanings we set out to inhabit and repudiate, the values we make efforts to live by or protest against, and the protest is also cultural. Culture resides primarily in the representations of the world exchanged, negotiated and, indeed, contested in a society.
>
> (Belsey, 2001: 7)

Again culture includes everyday practices and the objects in which a society's values are shaped, while the processes of negotiation and contestation do not privilege high culture over everyday life, or over the subcultures by which specific groups in a society identify themselves.

For Raymond Williams cultural and social histories are linked in incremental processes of change. Williams first bridged cultural and social histories in *Culture and Society* (1983, first published in 1958). Extending his argument in *The Long Revolution* he sees the creative mind as a catalyst to an incremental revolution in social attitudes beginning in changes in the institutions of culture and education. The process is non-teleological (has no given end) and produces new possibilities. In discussion of realism and the novel, Williams writes:

> The achievement of realism is a continual achievement of balance, and the ordinary absence of balance, in the forms of the contemporary novel, can be seen as both a warning and a challenge . . . any effort to achieve a contemporary balance will be complex and difficult, but the effort is necessary, a new realism is necessary, if we are to remain creative.
>
> (Williams, 1965: 316)

Culture, then, is a work of renewal in specific conditions. Those conditions are reproduced in cultural forms but may also be shown in new ways which eventually modify their reproduction.

I would compare Williams's model of cultural development with that of species evolution in Charles Darwin's *The Origin of Species* (first published in 1859) as reconsidered by Elizabeth Grosz in *The Nick of Time* (2004). Grosz argues that Darwin says very little about the origin of species and more of the descent of species, introducing an understanding of science 'which has no real units, no agreed upon boundaries or clear-cut objects' (Grosz, 2004: 21), which is in other words non-teleological. For Darwin, read by Grosz, the definition of a species is retrospective, not itself a factor in the course of evolution. Grosz continues:

> such a science could not take the ready-made or pregiven unity of individuals or classes for granted but had to understand how any provisional unity and cohesion derives from the oscillations and vacillations of difference.
>
> (Grosz, 2004: 21)

And she argues that differences between species can be read only comparatively:

> There is no origin of species because there is no unity from which descent is derived, only types, variations of differences and types of reproduction and descent.
>
> (Grosz, 2004: 25)

Difference is produced in language while the mutability of language itself is taken by Darwin as a model for evolution. The genealogies of life forms are akin for Grosz to 'other historically bound systems – to economies, to cultural systems that are equally embedded in the movement of history' (Grosz, 2004: 27). I find this illuminating because it brings biology into the realm of the produced, and enables me to see cultural structures as mutating, retrospectively categorised systems so that taxonomies change in history through the agency of those who use them.

In *The Long Revolution* Williams sketches three meanings for culture:

> There is, first, the 'ideal', in which culture is a state or process of human perfection, in terms of certain absolute values . . . second, there is the 'documentary', in which culture is the body of intellectual and imaginative work, in which . . . human thought and experience are variously recorded . . . third, there is the 'social' definition of culture, in which culture is a description of a particular way of life, which expresses certain meanings and values not only in art and learning but also in institutions and ordinary behaviour.
>
> (Williams, 1965: 57)

He does not see these as absolute, since all evolve and overlap. Culture as a way of life is studied to identify the meanings inherent in certain practices, ritualised or everyday, and how they change. This is like the Marxist concept of praxis – the gaining of appropriate understandings of present and past conditions to allow insights into possibilities for change – but has implications beyond Marxist histories. For me it is central to an idea of human agency, if constrained by institutional structures, contingent on webs of enabling and disabling factors, and a product of conditions, and linked to Williams's idea of incremental revolution through culture and education.

Williams writes that culture is one of the 'most complicated words in the English language' (1976: 76) and sees the multiplicity of meanings not so much as a disadvantage preventing a clear definition 'but as a genuine complexity, corresponding to real elements in experience' (Williams, 1965: 59). He argues that cultural works cannot be seen only as by-products of a social trend but are contributors to such trends (Williams, 1965: 60), and adds three levels to his three definitions:

> There is the lived culture of a particular time and place, only fully accessible to those living in that time and place. There is the recorded culture . . . from art to the most everyday facts: the culture of a period. There is also, as the factor connecting lived culture and period culture, the culture of the selective tradition.
>
> (Williams, 1965: 66)

Verb and noun

In *Keywords* (1976), at first conceived as an appendix to *Culture and Society*, Williams develops the definition of culture sketched in *The Long Revolution*. The length of the entry reflects its complexity – more than five pages compared to four for democracy, three for civilisation and one for modern, though class has nine. He notes the root of the word in the Latin *colere*, to cultivate, care for, protect or honour. *Colere* also means to inhabit, but the key meaning of *cultura* is a tending of natural growth and animal husbandry:

> **Culture** in all its early uses was a noun of process: the tending of something, basically crops or animals. The subsidiary *coulter* – ploughshare, had travelled by a different linguistic route, from *culter*, [Latin] – ploughshare, *culter*, [old English], to the variant English spellings *culter*, *colter*, *coulter*, and as late as [seventeenth-century English] *culture* . . . This provided a further basis for the important next stage of meaning, by metaphor.
>
> (Williams, 1976: 77, emphasis added)

Williams explains that the tending of natural growth was extended as metaphor to the cultivation of human intelligence:

> At various points in this development two crucial changes occurred: first, a degree of habituation to the metaphor, which made the sense of human tending

direct; second, an extension of particular processes to a general process, which the word could abstractly carry. It is of course from the latter development that the independent noun **culture** began its complicated modern history.

(Williams, 1976: 77, emphasis added)

Culture thus moves from verb to noun. Williams remarks in *The Long Revolution* that 'the cultural tradition can be seen as a continual selection and re-selection of ancestors' (Williams, 1965: 69); and in *Keywords* cites John Milton's *The Readie and Easie Way to Establish a Free Commonwealth* (1660): 'spread much more Knowledg and Civility, yea, Religion, through all parts of the Land, by communicating the natural heat of Government and Culture more distributively to all extreme parts' (Milton, 1660, cited in Williams, 1976: 78). He draws a distinction between culture's status as an independent noun in English and its use in French where it requires an object (the matter being cultivated). He writes, too, of the migration of the word into German, first as *Cultur* but by the nineteenth century as *Kultur*:

> Its main use was still as a synonym for *civilization*: first in the abstract sense of a general process of becoming 'civilized' or 'cultivated'; second, in the sense which had already been established for *civilization* by the historians of the enlightenment, as a description of the secular process of human development. There was then a decisive change of use . . . [when it became] necessary . . . to speak of 'culture' in the plural.
>
> (Williams, 1976: 78–79)

This parallels the growth of anthropology and ethnography as academic studies (no longer amateur activities of collecting) in which the ways of life of mainly non-European societies were compared. This meaning is encountered in Johann Gottfried von Herder's *Ideas on the Philosophy of the History of Mankind* (1784–91 – see Herder, 1969: 179–224), from which Williams quotes the following passage:

> Men [*sic*] of all the quarters of the globe, who have perished over the ages, you have not lived solely to manure the earth with your ashes, so that at the end of time your posterity should be made happy by European culture. The very thought of a superior European culture is a blatant insult to the majesty of Nature.
>
> (cited in Williams, 1976: 79)

Herder's view was radical in the 1790s in refusing the idea of a unified trajectory of human development of which European civilisation was the culmination, but it was not until the late twentieth century that ethnographers and anthropologists began to carry out fieldwork in Western, urban sites rather than in ex-colonial countries in Africa, Asia or Australasia.

Case study 2.1

From Octavio Paz, *The Labyrinth of Solitude*

The disparity of elements that can be observed in the Conquest does not obscure its clear historical unity. They all reflect the nature of the Spanish state, whose most notable characteristic was the fact that it was an artificial creation, a political construction in the strictest sense of the word. The Spanish monarchy was born from violence, the violence which the Catholic kings inflicted on the diversity of peoples and nations under their rule. Spanish unity was and still is the result of the political will of the state, which ignored the will of the elements that made it up . . . The speed with which the Spanish state assimilated and organized the conquests made by many individuals demonstrates that a single will, pursued with a certain coherent inflexibility, animated both the European and overseas undertakings. In a brief time the Spanish colonies achieved a complexity and perfection that contrast sharply with the slow development of those founded by other countries. The previous existence of mature and stable societies undoubtedly facilitated the task of the Spaniards, but the Spanish will to create a world in its own image was also evident. In 1604, less than a century after the fall of Tenochtitlan, Balbuena told the world of the *Grandeza Mexicana*.

The Conquest, then, whether considered from the native or the Spanish point of view, must be judged as an expression of a will to unity. Despite the contradictions that make it up, it was an historical act intended to create unity out of the cultural and political plurality of the pre-Cortesian world . . . If Mexico was born in the sixteenth century, we must agree that it was the child of a double violence, imperial and unifying: that of the Aztecs and that of the Spaniards.

(Paz, 1985: 91)

Comment

The undercurrent of this text is that the cultures of a newly dominant colonial power and a dominated colonised people (previously colonisers of other peoples in central America) are not so different as to prevent elements from each permeating the other. A similar point is elaborated by Setha Low (2000) in a study of the derivation of the urban form of the *plaza mayor*, or grid with central square for ceremonial and commercial purposes (which I consider in Chapter 9). Both indigenous and colonising societies were, as Paz observes, impregnated with religion (Paz, 1985: 84); for the peoples under

Aztec rule the arrival of the Spaniards seemed at first almost a form of liberation, then of forcing into another unifying state. The new cities laid out on grid plans in the New World may have been newer, then, to the Spaniards (whose medieval cities were labyrinths of winding alleys) than to the Aztecs. The form of the city established in the Law of the Indies – a set of planning directives by Phillip II from 1509 onwards, collected and published in a unified form in 1537 – is a fusion of cultural frameworks. Its nearest precedent in Spain is Santa Fe (see Chapter 9) completed in 1491, though Low mentions Islamic gardens as a possible influence (Low, 2000: 95). Tenochtitalan was already a grid-plan city with a large, open central plaza – before its destruction and rebuilding on the same site (now Mexico City). Madrid, too, has a *plaza mayor* which postdates the Conquest, possibly a design derived from that of Tenochtitlan. A meaning is inherent in it of cosmic order, or a masculinisation of space by a society whose elite worship a sky god (in the old and new worlds alike). Paz brings out the importance in such processes of cultural forms such as the grid, and institutionalisations as in the Laws of the Indies. Looking back, what is perhaps frightening is the continuity with which a desire for unity is expressed, and always defeated because it cannot correspond to the actualities it marshals into conformity to its image, or represses as deviant. Today Mexico is dominated by globalised capital based for the most part in its northern neighbour (Martin, 1999). While migrants take insecure, low-paid jobs in the affluent world Mexico City becomes for Jennifer Robinson, citing Saskia Sassen (1991), one of 'a select group of cities, some in poorer countries, . . . now deemed to have "global city functions" although they fall short of being first-order global cities' (Robinson, 2006: 99 – see case study 1.2). There are other histories. On 4 December 1914 Emiliano Zapata and Francisco Villa met at Xochimilco to draw up a treaty following several years of revolution and upheaval, holding a joint procession through Mexico City. The alliance fell apart but the Zapatista movement imagined a Mexico of communitarian social formation at village level – land was redistributed and 'Everything was taken in hand by the populace directly: including the railways . . . 1915 was one of the freest years the people of Mexico had ever known' (Fremion, 2002: 114).

Williams concludes in *Keynotes* that there are three main meanings for culture:

> (i) the independent and abstract noun which describes a general process of intellectual, spiritual and aesthetic development . . . (ii) the independent noun, whether used generally or specifically, which indicates a particular way of life, whether of a people, a period or a group . . . (iii) the independent and abstract noun which describes the works and practices of intellectual and especially artistic activity. This seems often now the most widespread use: **culture** is music, literature, painting and sculpture, theatre and film.
>
> (Williams, 1976: 80, emphasis added)

These correspond to his earlier definitions in *The Long Revolution* but in a different order: first is the cultivation of absolute values (now more abstract, as a process of seemingly autonomous intellectual development); second was a body of work (transposed as the third meaning in *Keywords*); and third was a way of life (now the second meaning). Williams's purpose is less to fix these meanings than to dwell on their overlaps:

> The complex of senses indicates a complex argument about the relations between general human development and a particular way of life, and between both and the works and practices of art and intelligence. Within the complex argument there are fundamentally opposed as well as overlapping positions.
>
> (Williams, 1976: 81)

Keeping in mind the issue of a national culture and the idea that cultural production is a site of contestation, I turn now to a case of a cultural institution in which nuances of the term *culture* overlap.

Tate

A context

Tate Modern is a museum of international modern and contemporary art, part of a national provision including Tate Britain at Millbank in London (housing the British collection), Tate of the North in a redundant warehouse in Liverpool, and Tate St Ives in a new building in a Cornish fishing village in which several modern British artists worked. Tate Modern is an internationally recognised cultural institution which plays a major role in defining modern art through the way it collects and exhibits particular works (and not others), the archetype of such institutions in its design and hegemonic function being the Museum of Modern Art, New York (Grunenberg, 1994).

Tate Modern occupies a 1950s power station on the south bank of the Thames. In its 1990s redesign the architectural practice Herzog and de Meuron retained the turbine hall as an open space the height of the building, with exhibition rooms at several levels accessed by escalators allowing visitors to experience the space of the building and sight of other visitors moving through it. Tate Modern has relocated London's cultural centre from the West End to a point at one end of a roughly north–south axis from its site to the financial district and the arts district of Hoxton. It is the success story of New Labour's rebranding of Britain as Cool Britannia following its election victory in 1997, and retains free entry. Cool Britannia did not denote a shift of cultural policy, however, but followed a decade in which flagship cultural institutions and cultural quarters (Bell and Jayne, 2004)

Figure 2.1
Tate Modern
(photo
M. Miles)

drove economic renewal (Bianchini et al., 1988; Arts Council, 1989). Tate Modern began like Cool Britannia's other flagship, the Millennium Dome, under the previous administration within a trend formalised as the creative city (Landry and Bianchini, 1995).

I find the idea of the creative city problematic on two grounds: there is little evidence to support its efficacy in delivering public benefit (Selwood, 1995); and its economic benefits are unevenly distributed while it has little to offer zones of deprivation (Evans, 2004). Culturally led redevelopment continues to be a commonly adopted strategy for urban renewal, nonetheless, as flagship cultural institutions in redundant industrial buildings contribute to reshaping a city's image. Cases include the Caixa Foundation's contemporary art gallery in an 1890s textile factory in Barcelona, art museums in disused railway stations in the Musée d'Orsay in Paris and Hamburg Station in Berlin, the Baltic art gallery in a disused flour mill in Gateshead, an arts centre in a disused cable factory in Helsinki, artists' studios in a custard factory in Birmingham, and the Warhol Museum in an early twentieth-century office building in downtown Pittsburgh.

For the London Borough of Southwark, one of the poorest in London, public investment in Tate Modern was seen as levering private-sector investment in adjacent commercial and residential spaces. Paul Teedon views design as central to Southwark's new image following acceptance that its industrial past was over, and sees the debate on design around Tate Modern as spreading to other projects nationally. He observes, 'a general ethos has been created that architecture and design, which pays close attention to the production of elements of high aesthetic quality, is becoming the norm and expected' (Teedon, 2002: 50). Most of the

Case study 2.2

From Samuel Mockbee, 'The Rural Studio'

I do not believe that courage has gone out of the profession [of architecture], but we tend to be narrow in the scope of our thinking and underestimate our natural capacity to be subversive . . . In other words, the more we practise, the more restricted we become in our critical thinking and our lifestyles. Critical thought requires looking beyond architecture towards an enhanced under-standing of the whole to which it belongs. Accordingly, the role of architecture should be placed in relation to other issues of education, healthcare, trans-portation, recreation, law enforcement, employment, the environment, the collective community that impacts on the lives of both the rich and the poor.

The political and environmental needs of our day require taking subversive leadership as well as an awareness that where you are, how you got there, and why you are still there, are more important than you think they are.

Architecture, more than any other art form, is a social art and must rest on the social and cultural base of its time and place. For those of us who design and build, we must do so with an awareness of a more socially responsive architecture . . . [which] not only requires participation in the profession but . . . also . . . civic engagement. As a social art, architecture must be made where it is and out of what exists there . . .

We must use the opportunity to survey our own backyards either to see what makes them special, individual and beautiful, or to note the unjust power of the status quo, or the indifference of the religious or intellectual community in dealing with social complacency or the nurturing of the environment.

(Mockbee, 1998: 73–74)

Comment

Samuel Mockbee died in December 2001, a few years after writing this account of his Rural Studio at Auburn University, Hale County, Alabama, a key task of which was the design and construction of rural housing. At the beginning of his narrative he recalls being at an architecture conference at which he and Michael Hopkins spoke in 1997. Hopkins was working at the time for Elizabeth Windsor, one of the richest women in the world, while Mockbee was working on a house for Shepard Bryant, 'the poorest man in the world' (Mockbee, 1998: 73), and his wife Alberta in Mason's Bend,

Alabama as a charity project. Asked from the floor how architects dealt with issues of social division Hopkins replied that perhaps architects should not make those kinds of decisions. Mockbee's essay is an extended form of his answer, that architects cannot wait for political change before intervening to make changes in areas such as housing provision. I have some issues with what he writes, particularly in that Mockbee seems to regard architecture as akin to exercise of a special insight – or second site as he says – and architects as obligated to use this privileged insight as agents of change. It is not that I detract, and see my own role as an academic within a university as facilitating critical attitudes to the society in which the university is situated; but Mockbee replicates the traditional avant-garde position of a deliverer, and I question the basis of this in as much as it might ignore the creative imaginations of the poor people who are Mockbee's clients. Yet the evidence of Mockbee's architecture suggests he was well aware of a need to hand over power to those whose rural housing he designed and built using appropriate technologies in keeping with local social and material conditions.

It could be asked why I include this case relating to small-scale settlements in a book on cities and cultures. I do so for two reasons: it draws attention to an example of a highly articulate form of grass-roots culture which challenges the dominance of the high-life architecture represented by Tate Modern; and because the process of making houses in appropriate materials and technologies is open to extension outside poor, rural neighbourhoods to inner-city districts and the housing of middle-income individuals and families as well. Mockbee's approach, employing his knowledge of architecture to facilitate translation of the ordinary citizen's needs into innovative housing, and working with teams of students as well as clients while using salvaged and recycled materials, could be compared with Hassan Fathy's (see case study 1.3) as a genuinely popular architecture, or as an alternative (and utopian) modernism. But Mockbee's work could be read as either postmodern or traditional, too, and his social aims – addressed through a consciousness of place, and that in turn through the people who live there at a particular time, the conditions in which they live and the material realities and scales of the place – remain the key to his practice. Since Mockbee's death, the Studio has moved to 'larger, more programmatically and technically sophisticated buildings' (Dean and Hursley, 2002: 11), and established an outreach programme for non-architecture students from other higher education institutions in North America.

design teams involved were small, young practices, for example Muf (a group of feminist architects who redesigned a nearby section of streetscape). Teedon argues that holding an international competition for Tate Modern gained recognition from 'a large proportion of the cultural elite' (Teedon, 2002: 53). Visitor numbers exceeded estimates – 35,000 on the opening day and a million in the first six weeks (Glancey, 2000; Jones, 2000, both cited in Teedon, 2002: 53) in contrast to the Dome.

Yet the Dome was an overt attempt to construct a national culture in a series of tableaux depicting national life, faith, technology and culture. Like the Great Exhibition of 1851 it occupied a signature building designed by Richard Rogers. Perhaps its failure is due, leaving aside inept management, to its timing, coinciding with a date on one of several calenders used in the world but at a point of British deindustrialisation. Its search for national cultural and historical narratives was thus contrived in the absence of a national success story after the loss of empire, while public services were in disarray and the gap between rich and poor again widening.

Tate as cultural space

Tate Modern is a new kind of cultural space, as much an exhibit as the art it houses and a new kind of cultural site. For George Cochrane, Tate Modern's Bankside Development Officer, it is 'a national and international landmark that has far reaching implications . . . for the city across the river' (Cochrane, 2000: 8). Esther Leslie writes that 'Tate Modern remakes the space of cultural encounter . . . is not just trendy, but in the vanguard of a reinvention of cultural spaces worldwide' (Leslie, 2001: 2–3). It does so, she adds, by replacing a coding of culture as preserve of an elite with another not of democracy but of creative commerce. Its industrial building lends Tate Modern an everyday aspect which departs from the veneration required by the temple-like museum buildings of the nineteenth century but, for Leslie, does so to serve the formation of consumer culture.

Tate Modern is progressive in maintaining free access to its collection and in initiating a training programme to help local people apply for jobs, though mainly in the low-paid areas of security and catering (Cochrane, 2000); but it is also a place in which to be seen and to consume, or be seen to consume in its café, restaurant and book shop, while offering a public education through its collection. That collection has been re-presented by separation of the international modern works from those of the British collection at Millbank, and perhaps more significantly by a new taxonomy in its first rehang: no longer categorisation by style, or by period and national school, but by theme, grouping works of disparate periods and styles by genres of subject matter. This, for Diarmuid Costello, restored an eighteenth-century taxonomy of 'discrete ensembles designed to trigger aesthetic insights by dint of novel association' (Costello, 2000: 24). Tate Modern has since had a second rehang though on similar lines of formal juxtaposition.

For me Tate Modern continues the traditions of the Tate at Millbank as a site of public improvement, and introduces a space for identity formation through consumption (Shields, 1992; Nava, 1992; see Chapter 8). For Costello, 'whatever else it may be, [Tate Modern] is first and foremost a building . . . it opens a space in which something may happen – primarily that works of art, and the people who gather to see them, may themselves appear in a particular way' (Costello, 2000: 15–16). Nick Stanley writes that visitors' experiences of Tate Modern are of its architecture: 'the architectural experience has become at least confused or conflated with the museological' (Stanley, 2000: 43). Costello adds that 'one often encounters a late rush of *grand projets* at the moment that whatever it is they celebrate . . . is nearing collapse' (Costello, 2000: 25). He cites North American railway stations built when Henry Ford adopted mass automobile manufacture, and British buildings in India shortly before its independence. His conclusion is that grandiose buildings are signs of 'cultural decadence, the expression of the psychological need of an age or a culture to create "monuments" to stave off, or mask, anachronism or impending obsolescence' (Costello, 2000: 25).

If modern art museums are widespread as international modernism is replaced by a collapsing of the boundaries of art, media, commerce and fashion, I suggest that institutions not only present what is about to be outmoded but also construct a narrative of a specific past which is produced to defend a structure of values seen as under threat. A parallel is the production of folkloric culture in a nation unsure of its identity in Victorian England, or the reproduction of Merry England in half-timbered suburban houses during the 1930s. Tate Modern constructs an encounter with contemporary art-space at a time of increasing political cynicism, a widening gap between rich and poor, privatisation of public services, and instrumentalism in cultural policy. This relates to the conditions of Tate's success in three ways. First, there is the context of a cultural turn in urban policy in which the arts are used to address (or mask) problems in other policy areas. Second, the success of Tate follows a popularisation of high culture in broadcast media and the press (as in colour magazines in middle-class Sunday newspapers since the 1960s), most recently in the publicity and controversy surrounding the annual Turner Prize for contemporary British art. Third, there is the growth of gallery education and outreach programmes to widen access to museums in the tradition of initiatives such as the Workers' Educational Association (Willett, 1967: 140–142). These factors suggest parallels between Tate Modern and the first Tate at Millbank.

Case study 2.3

Skinningrove, Cleveland, UK

Figure 2.2 *Pigeon lofts at Skinningrove, Cleveland, 2001 (photo M. Miles)*

Comment

Skinningrove is an industrial village on the Cleveland coast in the north-east of England, built to house workers in the mining and steel industries. It once had a cricket ground. All the pits have closed, and the nearest steel works employs a small, highly specialised labour force from elsewhere. There is a local school, a shop and post office, a pub and an infrequent bus service to the nearest town with its supermarket and cinema. There is not much to do in Skinningrove, and many local children and young people work in the informal economy or in illegal work such as poaching. For the millennium, villager Tommy Evans proposed a mass toilet flushing at midnight to reveal the leakage in the village sewer-pipe, and general exposure of the incompetence or indifference of officialdom to a place like Skinningrove. In June 2000 a flood devastated the village. Raw sewage ran through living rooms to a depth up to five feet, the driftwood on the beach (a normal source of fuel) was declared unsanitary and burnt, and the annual bonfire and firework display was cancelled. New flood defences have since been installed, though disregarding the advice of local people as to exactly where they are needed. Local and national authorities see Skinningrove as a blank on the map, written off in the Durham County Plan as a site of depopulation in which no further investment should be made – a policy designed to encourage emigration. Most inhabitants remain because they cannot afford to move to places where housing costs are higher (almost everywhere else), and because the village gives them a sense of belonging. Unemployed men have coffee mornings and participate in local politics. Some have pigeon lofts on the adjacent hillsides (Fig. 2.2) – self-build structures using found materials. They house collections of shells or driftwood found on the beach, often with an old armchair. One or two have stoves for smoking fish, and many have stores of found materials for future building and adaptation. In most cases the owners can no longer afford pigeons, originally kept for racing (on which the betting was taken over by outside parties). The lofts remain as material evidence of a culture – 'a particular way of life, which expresses certain meanings and values . . . in institutions and ordinary behaviour' (Williams, 1965: 57). There was an attempt to establish an art project in Skinningrove in 2000 but – leaving aside the intricacies of funding and tactics – I wonder if it failed because the local culture is so strong and visible that there is little an artist can do. Art cannot compete for meaning with the material culture of the lofts or the aesthetics of the found objects housed in them. An artist's efforts at ethnography might produce interesting insights but the informal economy breeds skills of improvisation, adaptation and resourcefulness required for survival. The waterfront at Skinningrove is called Beirut on account of the number of boarded-up houses.

The Tate at Millbank

The first Tate was a key element in the 1890s redevelopment of Millbank for an artisan class; it educated visitors in the interpretation of art, specifically offering a narrative of British art to equate with that of other European powers and trade competitors. The Tate Gallery, as it was known prior to Tate Modern, opened in 1897 and was a gift to the nation from the philanthropist Henry Tate, who made his money from the invention of the sugar cube (and the sugar trade, a key user of slaves in the eighteenth and early nineteenth centuries). A contemporary sketch in *Penny Illustrated Paper* (Taylor, 1994: 21) shows sailing boats on the Thames in front of it, like a prototype of the waterside views used a century later to promote London's docklands (Bird, 1993). According to Brandon Taylor, Tate's public was mainly middle-class but had been intended to be more varied. The caption to an illustration of spectators inside Tate lists the classes who compose its public:

> The well-to-do citizen . . . and the citizen who having but doubtful domicile went in to have a rest; the dainty maiden, the costermonger, and the old maid; school board children, foreigners, the fashionable curate with his sister or somebody else's sister, the nondescript individual who may have passed a few days or longer in the original building.
> (*The Daily Graphic*, 24 August 1897, cited in Taylor, 1994: 23)

Another report mentions working people 'wandering' the rooms and 'gazing' at the art (*The Norwood Press*, 21 August 1897, cited in Taylor, 1994: 22) as if lost. Taylor notes an inconsistency between such verbal accounts and illustrations of Tate like that cited above, and sees the idea of a mixed audience as fantasy (1994: 23). He adds, 'I think we can be sure that the desire of the new art public was to be free of the contamination of the lower classes altogether' (Taylor, 1994: 24). Nonetheless, Tate's building was, like Tate Modern now, situated close to an area of lower-class housing; and the idea of drawing in a public from such sites has its own history, not least in Matthew Arnold's concept of culture as a means to social cohesion in a period of religious decline. In the 1890s, and perhaps the 1990s, urban improvement as in Tate's transformation of its marshy site, previously occupied by a prison, goes in hand with social improvement as artisans and working-class people gain access to middle-class culture, in a tradition of reform designed to prevent insurrection while retaining the core values of the establishment.

Tate promoted British art in competition with French art but was also a model of social exchange in which difference is assimilated as antidote to insurrection (seen as another French trait). One response to the threat of insurrection was provision of better living conditions for the poor, or the deserving poor, in a period which also saw a rise of efforts to end reliance on drink (a defence against pain or misery) and idleness; another was to extend cultural access while constructing a national cultural identity. Taylor summarises:

> The founding of the gallery also raised the possibility that 'philanthropic' individualism was ultimately compatible with a cultural authority deriving from the state and its civil service, and that something identifiable as 'the public good' could be defined by the conjuncture of the two . . . The Tate Gallery also represented a relatively new – for 1897 – compromise between the interests of entrepreneurial capitalism and the amateurish style of cultural management typical of the old nobility.
>
> (Taylor, 1994: 27)

This has an uncanny resemblance, with a little modification, to what could be written of Tate Modern – the alliance of capital and state, the informality of taste and the edge of money, and the idea that the public good is served by widening access to culture. It alludes obliquely to the type of person who spent time in the original building on the site in the *Daily Graphic* quote (above) – the Millbank Penitentiary, a structure of six polygons built between 1813 and 1816. There are clearly several histories of Tate, but I want to end with this one: the transition from a site of discipline to one of social improvement.

Taylor describes Millbank as 'heavily associated with dirt, inaccessibility and crime' (Taylor, 1994: 12), and the penitentiary as a depot for convicts awaiting transportation as well as a women's prison and military prison. The nineteenth century saw efforts to cleanse cities, like London with its stinking Thames, of dirt and disease, in a longer development from the seventeenth century of purging urban spaces of, first, the insane and vagrant (Foucault, 1967), then the miasma of the dead (Illich, 1986), and by the nineteenth century the smell of the poor (Corbin, 1996). The reforms were progressive and beneficial but their ideological basis privileged the entrepreneurial class as owners of clean bodies and inheritors of cultural knowledge. The Tate was a reclamation of its site from 'the ravages of darkness and sin' (Taylor, 1994: 13) which fits it well in this trajectory. Illustrations of designs through the 1890s (Taylor, 1994: 15) mix classical vocabularies in the *beaux-arts* style, and the final building has the look of a temple of art. Taylor summarises:

> The particular form given to this metaphor of enlightenment relied massively upon the demise of the old prison, combined with the appearance of the new gallery in its guise as harbinger of the 'progress of civilization'. It was a network of associations that was regularly endorsed.
>
> (Taylor, 1994: 16)

The prison was demolished in 1892. The narrative of urban improvement was underpinned by a view that culture, associated with universal value, is a social glue. The establishment of an Arts Council in the 1940s to 'increase the accessibility of the fine arts . . . [and] improve the standard of execution of the fine arts' (in Willett, 1967: 198) is a logical step in this pathway.

The kinds of art displayed in Tate Modern are not those which appeared in the first Tate, but the model of improvement remains. John Willett writes that, in the 1960s, government saw one of its roles as enabling the arts to flourish but he worries that standards are determined by 'the taste prevailing among dealers and critics and virtually the whole of the official art world' (Willett, 1967: 204). I wonder if a new instrumentalism by which the arts solve problems from crime to unemployment and social exclusion (Yúdice, 2003) is a continuation of a project for social cohesion. If so, culture as the arts begins to impact culture as a way of life, in the interests of a national culture, while cultural institutions displace the everyday cultures of the neighbourhoods they recode as cultural sites. Leslie's conclusion is that Tate Modern's remoulding of the art institution recasts culture as an adjunct of commerce:

> Tate is a brand that niche-markets art experience. Its galleries are showrooms. However, this is still art and not just business. the commodity must not show too glossy a face. The reclamation of an industrial space that provides the shell for the Tate Modern lends the building a fashionably squatted aspect . . . unlike the old purpose-built Tate with its massive portico and reek of victory over the penitentiary that once stood in its place. Visitors ascend into that art temple . . . Tate Modern begins with a descent . . . home to the new-style 'accessibility rules' culture industry.
>
> (Leslie, 2001: 3)

Tate denotes social improvement as well. Whether it constructs a postmodern national culture is doubtful in that cities now compete globally independently of the nation-states in which they are sited. The juxtapositions of exhibits on a thematic basis refuses the trajectory model of anthropological collection like that donated by Lieutenant-General Pitt Rivers to Oxford University in 1883 (Coombes, 1994a: 119) to affirm national superiority. Yet when Annie Coombes writes of the 1890 Earls Court Exhibition, 'Time and history were reordered and remade in the image and likeness of an unstoppable British supremacy' (Coombes, 1994a: 85) I wonder if Tate Modern constructs the supremacy of a certain kind of culture and a certain cultural lifestyle.

Critical culture?

If culture is what is incorporated in the creative and cultural industries of which Tate Modern is an iconic presence, there is another definition of culture in cultural criticism (*Kulturkritik*), denoting critical engagement with cultural and social forms in specific histories. Francis Mulhern reads it as 'a critical, normally negative discourse on the emerging symbolic universe of capitalism, democracy and enlightenment – on the values of a condition and process of social life for which recent French coinage furnished the essential term: *civilization*' (Mulhern, 2000: xv, citing L. Febvre, 1973, *A New Kind of History*, London, Routledge, pp. 219–257). Mulhern associates Williams with this position: 'The second major European centre of *Kulturkritik* was England, whose counterpart tradition is the subject of Raymond Williams's classic study, *Culture and Society*' (Mulhern, 2000: xvi). He juxtaposes Herder's pluralism – 'Cultures were the symbolic forms of life of human groups, shaped in diverse conditions and growing into new shapes as they encountered new demands and opportunities' (Mulhern, 2000: xvi) – to the implication of a national culture in Herder's remark that 'Even the image of happiness changes with each condition and climate . . . each nation has its centre of gravity within itself, just as every sphere has its centre of gravity' (Herder, 1969: 185–186, in Mulhern, 2000: xvii). But Williams reminds me that there are dissenting as well as affirmative cultural movements. The long revolution is a process of change through adaptation of social institutions, as in education, and perhaps a grass-roots alternative to the hegemonic role of cultural institutions in reproducing narratives.

In conclusion, I return to culture as a means to intelligibility in a complex world, citing Belsey: 'The subject is what speaks or writes, and it does so in a language which is the inscription of certain knowledges – of a culture' but 'signification always precedes us as individuals, and meanings pertain at a specific historical moment. To speak or write is to reproduce, however differentially, the meanings we have learnt (Belsey, 2001: 10). As Belsey says, the protest is cultural, too.

Notes for further reading

Chapters 1 and 2

These notes relate to the general issues outlined in Chapters 1 and 2, and suggest avenues for more detailed study. I begin with a number of anthologies relevant to the terrain of urban cultures. Among these are *The City Reader* (LeGates and Stout, 3rd edition 2003) with sections on areas such as 'The Evolution of Cities', 'Urban Culture and Society', 'Urban Planning History and Visions' and 'The Future of the City'; *The City Cultures Reader* (Miles, Hall and Borden, 2nd edition 2003), with sections on 'Symbolic Economies', 'The Culture Industry', 'Cultures and Technologies' and 'Utopias and Dystopias'; and *Gender Space Architecture: An Interdisciplinary Introduction* (Rendell, Penner and Borden, 2000), which I regard as an exemplary book in the way it sets out and critically introduces its material, offering access to issues of gender and contemporary urban space in fairly short textual extracts. Edward Soja's *Postmetropolis* (2000) includes his account of Çatal Hüyük, and considerable material on recent and contemporary urban theory, with an emphasis on Los Angeles. A review of critical sources is given in his earlier *Postmodern Geographies* (1989).

Turning to individual contributions to current debate on cities and urban cultures, James Donald gives a readable and incisive commentary on the modern city, particularly as seen in mass culture, in *Imagining the Modern City* (1999). This could be read with the more eclectic *Images of the Street*, edited by Nicholas Fyfe (1998), with chapters on fascist Rome, homelessness, street-level erotics, the Indian street and surveillance. An essay on the layered city by Peter Marcuse is included in *The Urban Lifeworld: Formation, Perception, Representation* (Madsen and Plunz, 2002). *Small Cities*, edited by David Bell and Mark Jayne (2006), includes chapters on Port Louis, Mauritius; Singapore; Portland, Maine; and Weimar, among others. It helpfully makes a case that small cities generate particular kinds of conditions and dynamics, and could be read alongside Jennifer Robinson's *Ordinary Cities* (2006) with material on several African cities within a questioning of the preponderance of narratives of world or global cities in contemporary urban theory. That global cities are not unified entities is demonstrated, however, in *Living the Global City: Globalization and Local Process*, edited by John Eade (1997), which presents research on the fluid cultural formations of a district of south London. Anthony King's *Spaces of Global Cultures: Architecture Urbanism Identity* (2004) considers the socio-cultural effects of architectural globalisation. On cities as cultural sites, Peter Jukes's *A Shout in the Street: The Modern Cities* (1990) covers London, Paris, Leningrad and New York in the early modernist period (relevant to Chapter 3). At a more detailed level, Simon Sadler's *The Situationist City* (1999) charts the work of the Situationists in Paris in the 1960s and their link to Henri Lefebvre up to 1968; Jane Rendell's *The Pursuit of Pleasure: Gender, Space and Architecture in Regency London* (2002) is a critical-historical study of urban spatial practices; and David Pinder's *Visions of the City* (2005) is an account of urban utopianism and its interface with modernist art (also relevant to Chapter 3). Tim Cresswell's *In Place Out of Place: Geography, Ideology, and Transgression* (1996) gives an account of alternative urban cultural forms such as New York graffiti art; it can be read beside David Sibley's *Geographies of Exclusion* (1995) which links similar issues to mass culture and advertising as well as psychoanalysis and Alain Corbin's urban anthropology. *Histories of the Future*, edited by Daniel Rosenberg and Susan Harding (2005), reconsiders the growth industry of urban future imaginings, in mass culture as well as broad economic and political scenarios, with chapters on the cell phone, UFOs and timeliness as well as Futurism and Surrealism.

Although his style is daunting (owing to a fugue-like structure) Henri Lefebvre's *The Production of Space* (1991) and *Introduction to Modernity* (1995) are worth the effort, though my own reading was initially via key words in the index. I found Rob Shields's *Lefebvre, Love & Struggle* (1999) and Stuart Elden's *Understanding Henri Lefebvre* (2004) helpful. On cultural theory I have found Catherine Belsey's *Culture and the Real* (2005) invaluable in trying to understand post-structuralist approaches to culture, with her earlier *Shakespeare and the Loss of Eden* (2001). A number of theoretical issues are addressed in *The Problems of Modernity*, edited by Andrew Benjamin (1991), which includes Janet Wolff's essay on the absence of women in modern literature. Raymond Williams's writing remains available in paperback, and *Keywords* (1976) is probably the most useful place to begin, though I recommend *The Politics of Modernism* (1989) – a collection of essays on modernist art in its urban social and technological contexts – as well. Among more recent contributions is Iain Chambers's *Culture after Humanism* (2001), which reconsiders the Baroque, and how pasts are remembered. On urban planning and democracy in contemporary contexts I suggest *Cities for Citizens*, edited by Mike Douglass and John Friedmann (1998). Leonie Sandercock's *Towards Cosmopolis* (1998) gives a helpful and succinct history of planning models from the rational comprehensive to the radical, and argues for planning for multi-ethnic cities. These books are largely theoretical, and would be complemented by the practicalities described in *Rural Studio* by Andrea Oppenheimer Dean and Timothy Hursley (2002), an account of Samuel Mockbee's Rural Studio (see case study 2.2).

Finally, I draw readers' attention to two works by Stephen Barber, between urbanism and travel writing: *Fragments of the European City* (1995), on the transitional stages of cities; and *Extreme Europe* (2001) on his experience walking peripheries and peripheric spaces (or non-places) in Europe. Like the writing of Octavio Paz cited in case study 2.1, these demonstrate the relevance of non-academic literatures to academic study.

Part Two

Interactions

3 Cities producing culture

Learning objectives

- To examine the role of cities as centres of cultural production and innovation
- To consider the case of modernism in Paris
- To reconsider the relation of new movements in the arts to the ways of life of cultural producers such as artists

Introduction

In the last two chapters I reconsidered cities, culture and cultures. In this chapter I look at the emergence of new cultural movements in Paris as a distinctly modern city in the second half of the nineteenth century. I focus on how art movements are produced, and the idea that there is a way of life associated with cultural production. I cite Baudelaire's art criticism as self-consciously modern, asking first what constitutes the modern period. Looking at Paris in Baudelaire's time through the work of Walter Benjamin I find a utopian aspect in modernism. Bringing the argument nearer the present I reconsider Pierre Bourdieu's analysis of cultural formations (1993) and end with a note on the particular form of the artists' banquet. In case studies I cite Fernando Pessoa's writing in Lisbon in the early twentieth century, and an extract from a short book on allotment culture by David Crouch. Finally, I include a photograph of a men's public lavatory in Hull as evidence of everyday creativity.

Paris and modernity

The time and site of modernism

The modern period in art history begins in most library classifications in the 1880s, but the idea of a modern culture begins before that. Baudelaire wrote of 'The Painter of Modern Life' in 1859, and John Ruskin coincidentally used the term in *Modern Painters*, written between 1843 and 1860. Ruskin and Baudelaire saw the art of their time as expressing an awareness of living in a period that is different, a departure from and not a gradual adaptation of the past. Ruskin saw being modern as accepting the place of the natural sciences in knowledge, argued for the establishment of a School in Natural Science at Oxford University and contributed to the design of its decoration. The interior featured a glass and cast iron roof begun in 1855. Under its second version – the first collapsed – the debate on Darwin's theory of natural selection took place in 1860 (Fuller, 1988: 93–99). Ruskin was later disillusioned with this attempt at a scientific gothic style but the tension between the gothic and the scientific was itself modern, premised on rationality, technology and a revival of medievalism which, rather than being a regression to a twilight zone of history, offered the model of co-operation in guilds as a specifically city-based model for nineteenth-century social reform.

Baudelaire's art criticism was conditioned by technologies but also by emerging class divisions as Paris was transformed into a bourgeois city around him. A motif of this period is the glass-roofed arcade – the technology used in the natural science building in Oxford. From the 1820s the arcades cut passages through the city blocks of Paris, offering refuge for strollers from the rush of horse-drawn carriages in narrow streets. The strollers observed each other as much as the luxury goods on display in the arcades, while railways made Paris the commercial centre of a region, creating a critical mass of buyers to sustain luxury trades. International expositions in Paris in 1855, 1867, 1878 and 1889 added to the impression of a capital of innovation, as did the remodelling of the city centre by Baron Haussmann. But the new boulevards also cut through the city's working-class quarters to inscribe more than a new design, as John Rennie Short comments:

> The winners were the moneylenders, who made profit from extending credit to the government, the state itself, which pushed broad, easily policed streets through pockets of working-class resistance, and the bourgeoisie of Paris, who saw the city written in their own image and in their own interest.
>
> (Short, 1996: 177)

Walter Benjamin writes that 'The true goal of Haussmann's projects was to secure the city against civil war . . . to make the erection of barricades . . . impossible for all time' (Benjamin, 1999: 23). In fact barricades were set up in the Commune in 1871, but the Commune was defeated and with it the radicalism of artists such as Gustave Courbet. What remained was a city of boulevards, waterways, parks, bars, cafés and picnic sites painted by the Impressionists in the 1870s (their first exhibition taking place in 1874). The art and literature of the 1880s and 1890s then turned to the mental state of the artist or writer, which can be taken as the beginning of a move to abstraction and autonomy from both natural appearances and social institutions defining modernist art from German Expressionism in the 1910s and 1920s to Abstract Expressionism in New York in the 1940s and 1950s. It is unhelpful to draw a line as the end of one period and the beginning of the next, not least because art continues to be produced and collected after the moment of innovation itself, but broadly a century of modernism in the arts connects Baudelaire to the abstract art of the postwar years. The point about this digression into art history, here, is that the whole of this development depended on a critical mass of artists, writers, critics, patrons and intermediaries in large cities. Modernism took place, that is, in the European metropolis, and it is difficult to see how the pace of innovation could have occurred without the support of that critical mass – and the publics which a large city brought to the arts just as it brought them to luxury goods in the arcades.

Painting modern life

Modernism requires a new kind of artist, or critic. Baudelaire's essay 'The Painter of Modern Life' concerns the work of artist-illustrator Constantine Guys but is equally an exposition of how a modern critic writes – as much about Baudelaire as cultural intermediary as about Guys – and takes in the ambience of a contemporary, metropolitan city. Baudelaire categorises Guys as 'not precisely an *artist*, but rather a *man of the world*' who is in his element in the crowd (Baudelaire, 1998 [1863]: 495–496). Guys is a *flâneur* (male stroller) who observes other people in his wandering observation of the city. For Janet Wolff (1991) the absence of women as protagonists and narrators in this genre is evidence of a masculine cultural history. Elizabeth Wilson comments that renewed interest in the *flâneur* on the parts of male writers in sociology and cultural studies in the 1980s followed new interests in consumption and tourism; and that recent commentators are ambiguous as to whether the *flâneur* is confined to the nineteenth century. Wilson argues that the success of the literary genre which was spawned by *flâneurie* depended on a new kind of journalism. She adds of the *flâneur*:

Case study 3.1

From Fernando Pessoa, *The Book of Disquiet* (*Libro do Desassossego*)

The wind was rising . . . First it was like the voice of a vacuum, a sucking of space into a hole, an absence in the air's silence. Then there was a sobbing, a sobbing from the world's depths, the realization that the panes were rattling and that it really was the wind. Then it sounded louder, a deafening howl, a disembodied weeping before the deepening night, a screeching of things, a falling of fragments, an atom from the end of the world.

And then it seemed . . .

(Pessoa, 2001: 52, ¶ 52, lacunae in original)

O night in which the stars feign light, O night that alone is the size of the Universe, make me, body and soul, part of your body, so that – being mere darkness – I'll lose myself and become night as well, without any dreams as stars within me, nor a hoped-for sun shining with the future.

(Pessoa, 2001: 241, ¶ 280)

What I most of all feel is weariness, and the disquiet that is its twin when the weariness has no reason but to exist. I dread the gestures I have to make and am intellectually shy about the words I have to speak. Everything strikes me in advance as futile.

The unbearable tedium of all these faces, silly with intelligence or without it, nauseatingly grotesque in their happiness or unhappiness, hideous because they exist, an alien tide of living things that don't concern me . . .

(Pessoa, 2001: 183, ¶ 337, lacunae in original)

Comment

There is a bronze statue of Pessoa outside Café Brasiliero in the centre of Lisbon's Chiado district (fig. 3.1). Tourists like to be photographed in the bronze empty chair next to him.

Figure 3.1 *Lisbon, statue of Pessoa outside Café Brasiliero (photo M. Miles)*

He was a customer here and his urban wanderings are set in the Baixa district below – the grid-plan district rebuilt after the earthquake of 1755. But Pessoa adopts multiple personalities:

Some of Pessoa's most memorable work in Portuguese was attributed to the three main poetic heteronyms – Alberto Caeiro, Ricardo Reis and Alvaro de Campos – and to the 'semi-heteronym' called Bernardo Soares, while his vast output of English poetry and prose was in large part credited to heteronyms Alexander Search and Charles Robert Anon, and his writings in French to the lonely Jean Seul.

(Richard Zenith, introducing Pessoa, 2001: viii)

The Book of Disquiet, as published, is an edited version of work left after his death in 1935 (spending all his life in Lisbon except a spell as a child in Durban, South Africa, where his stepfather was Portuguese consul). More than half its contents are from the last six years of his life, much of that under the name Bernardo Soares. It contains aphorisms, reflections on the life of a clerk, engineer or other persona, and passages of description. The most frequent locale is Rua dos Douradores where Soares lived, worked in an office and dined in an upstairs restaurant. In passage 134, he states: 'I seek and don't find myself. I belong to chrysanthemum hours, neatly lined up in flowerpots' (Pessoa, 2001: 121). He writes of rain, and dim lights, snatches of dream seen on a wall in half-sleep, time dragging, absurdity – a vision more fragmentary than Baudelaire's, that lacks the utopian content discerned by Benjamin. But Lisbon, like Paris and London, is an imperial city. Pessoa wanders its streets but is distanced from it by his dozens of heteronyms – espousing Baudelaire's negativity in his third preface to *Fleurs du mal* (see main text). Pessoa's book ends with an image of the author's death as 'one less passer-by on the everyday streets of some city or other' (Pessoa, 2001: 392). The disquiet of this observer whose inner life at times verges on insanity could be compared with the anxiety of Edward Munch's paintings and prints in the 1890s, or Schoenberg's music. It seems required.

It is still uncertain whether she or he is simply strolling, loitering and looking . . . or whether these activities must be transformed into a representation – journalism, film, novel – in order to qualify as *flânerie*. Women are especially caught in this ambiguity. When I suggested that nineteenth-century prostitutes might be considered as the *flâneuses* of their time I was being (intentionally) provocative, and one telling argument against this view is that prostitutes are not strolling and observing, but are *working*. Some feminists have taken this argument further and pointed out that even as shoppers women are working.

(Wilson, 2001: 91)

Wilson concludes that ambiguities remain. But is the *flâneur*'s gaze ephemeral or furtive? Baudelaire, writing on Guys, refers to 'that quality which you must allow me to call "modernity"' as 'the ephemeral, the fugitive, the contingent, the half of art whose other half is the eternal and the immutable' (Baudelaire, 1998 [1863]: 497). It appears, whatever the ambiguities, that the idea of the artist or writer as articulating personal experiences of being in a crowd, or in the ordinary places of urban living, is as specific to the conditions of a large city as it is gender-specific. Those crowds came together only in cities, just as cities offered the critical mass of buyers to sustain markets in luxury goods in the arcades and, with those markets, a wider and larger public of onlookers. If, thinking back to Chapter 1, proximity is a defining condition of a city and creates the interstices in which informal sociation tends to take place as well as anonymity in a crowd, then the conditions of a metropolitan city are framed also by a new kind of culture in mass interest in consumption and consumers, and the daily busy-ness, so to speak, of watching others whose anonymity confirms that of the watcher. This has a fugitive aspect as much for those who watch others watching them as for those objectified by the gaze.

In the self-consciously modern city, a division occurs between a public realm of streets, squares and monuments, and a private – or domestic – realm of the household and ordinary life. Citing Naomi Schor (1992), Ben Highmore notes a feminisation of everyday life 'carried out within the domestic sphere tradition-ally presided over by women' (Schor, 1992: 188, in Highmore, 2002a: 15) and distinct from a masculinised public realm. The aim for men, in the accounts of Mike Featherstone (1995) and Xiaobing Tang (2000) which Highmore cites, is to overcome the everyday world while inevitably dwelling in it. But the *flâneur* is uncomfortable with this, observing a life in which a vicarious pleasure is taken optically, at a distance from the life observed. The modernist view is a view of streets but strangely outside or as if above them at the same time. Richard Sennett's description of a walk from Greenwich Village to French restaurants in mid-town New York in *The Conscience of the Eye* (1990) has this quality as he observes drug deals, women in bars, equestrian shops used by leather fetishists, ethnic spice shops, a library, a neighbourhood of decaying gentility and dull apartment blocks:

> New York should be the ideal city of exposure to the outside. It grasps the imagination because it is a city of difference par excellence, a city collecting its population from all over the world. Yet it is here that the passion of the Parisian poet [Baudelaire] . . . seems contravened. By walking in the middle of New York one is immersed in the differences of this most diverse of cities, but precisely because the scenes are disengaged they seem unlikely to offer themselves as significant encounters.
>
> (Sennett, 1990: 128)

He never says what he eats.

I would compare Sennett's story with the following account of an imagined Paris in 2000 by Tony Moilin (written in 1869), cited by Walter Benjamin in his study of the arcades:

> The government had wanted the streets belonging to the people of Paris to surpass in magnificence the drawing rooms of the most powerful sovereigns . . . First thing in the morning, the street-galleries are turned over to attendants who air them out, sweep them carefully, brush, dust, and polish the furniture . . . Between nine and ten o'clock this cleaning is completed, and passersby . . . begin to appear . . . Entrance to the galleries is strictly forbidden to anyone who is dirty or to carriers of heavy loads; smoking and spitting are likewise prohibited here.
>
> (Tony Moilin, *Paris en l'an 2000*, 1869, in Benjamin, 1999: 54–55)

This could describe the Burlington Arcade in London (see Rendell, 2002) and many shopping malls catering for today's democratised consumption of items, the primary use of which is stating identity choices.

The arcades

Benjamin, wandering the arcades of Paris in the 1920s, notes a sign: 'Angela 2nd floor, to the right' (Benjamin, 1999: 40), and writes that in the arcades there is trade but no traffic as in the streets:

> The arcade is a street of lascivious commerce only; it is wholly adapted to arousing desires. Because in this street the juices slow to a standstill, the commodity proliferates along the margins and enters into fantastic combinations, like the tissue in tumours. – The flâneur sabotages the traffic. Moreover, he is no buyer. He is merchandise.
>
> (Benjamin, 1999: 42)

Graeme Gilloch, commenting on Benjamin's 'Arcades Project', writes that 'Commodity fetishism was the principal characteristic of the dreaming urban collectivity' (Gilloch, 1996: 118). Lithographs were among the goods on offer in an art market expanded to the middle classes who could not afford original paintings but could afford prints.

The arcades foregrounded a new architecture of steel, glass and polished surfaces, and were built at the same time as railway stations which used a similar technology to construct wide arches over platform spaces. Several such arcades were in an area of Paris between rue Croix-des-Petits-Champs, rue de la Grange-Batelière, rue Ventador and boulevard Sebastapol. The first, Passage Viollet and Passage des Deux Pavillions, were built in 1820, followed by Passage des Panoramas and Passage de l'Opéra in 1823 (illustrated, Benjamin, 1999: 49), Passage Véro-Dodat in 1823–26, and others to Passage de Prince in 1860. A few had outlets for one commodity – such as lithographs in Passage du Caire – or basic commodities – fish in Passage du Saumon – but most housed a mix of restaurants, music and fine print shops, wine merchants, hosiers, haberdashers, bootmakers, milliners, tailors, book shops and other outlets for non-necessities in one gas-lit enclosure. Only a minority of those who came to the arcades could afford to buy the goods displayed. Benjamin cites a journal of 1852:

> These arcades, a new invention of industrial luxury, are glass-covered marble-panelled corridors . . . whose proprietors have joined together for such enterprises. Lining both sides . . . are the most elegant shops . . . the arcade is a city, a world in miniature . . . in which customers will find everything they need.
>
> (in Benjamin, 1999: 31)

This suggests a world in miniature, an artificial world. It is here that Benjamin, looking in toy and curiosity shops in the 1920s when some of the arcades are semi-derelict and others demolished, finds a latent utopian consciousness which accompanies and complements the fetishism of the commodity.

Karl Marx wrote of commodities as betraying no trace of their production by workers, thereby taking on an (as if) magical quality (in 'Economic and Philosophical Manuscripts'), and for Benjamin mass-production extends this to new degrees of artificiality. To me there is an affinity between Benjamin's work and Baudelaire's in this respect, but the dystopian and utopian tend to be separated in Baudelaire's writing while the search for a redemptive utopian predominates in Benjamin's. It may be that cultural production bifurcates into unique commodities (art) and reproducible ones (prints and photographs), so that the former retain an aura in converse relation to the mass-production of the latter and as tacit resistance to commodities – though it is the aura of the artwork which is then marketed to differentiate art's high price from the low prices of reproducible cultural goods. Similarly, displays of commodities were replaced at regular intervals as fashions changed and an equivalent of art's aura shifts from mass-produced goods to the style in which they are produced. Featherstone sees newness as a key aspect of Baudelaire's idea of modern life while modern societies produce 'commodities, buildings, fashions, social types and cultural movements . . . destined to be rapidly replaced by others' to reinforce the transitoriness of the moment (Featherstone, 1995: 150). Again the metropolitan city is the ground for all this, the only site of a necessary critical mass of people for a privileging of innovation in architecture, fashion and art, as in the patterns of human sociation on which Georg Simmel writes.

Benjamin reconstructs the arcades as dreamworlds in which pasts and presents co-exist. James Donald remarks that Benjamin's adoption of Baudelaire 'as his surrogate' is explained by a fascination with 'the way that the displacements brought about by Haussmannization lent a fantastic and elusive quality to life in the city' (Donald, 1999: 47). Benjamin makes much of the iron and glass construction of the arcades: 'With iron, an artificial building material appeared for the first time in the history of architecture' (Benjamin, 1997: 158). He then writes that old and new images of production intermingle in a collective consciousness: 'These images are ideals, and in them the collective seeks not only to transfigure, but also to transcend, the immaturity of the social product and the deficiencies of the social order of production' (Benjamin, 1997: 159). A desire to break with what is outdated leads to a projection of the ideal on to a primal past: 'In the dream in which every epoch sees in images the epoch which is to succeed it, the latter appears coupled with elements of prehistory – that is to say of a classless society' (Benjamin, 1997: 159; see also Wolin, 1994: 175–176). The arcade is the site of an intensive commodification and for Benjamin a potential consciousness of liberation glimpsed in everyday things.

The display of goods in the arcades was a prototype of the display of modern art in museums designed specifically for it, where techniques of display were borrowed from retail practices – as in the Museum of Modern Art, New York (Grunenberg, 1994) – and clusters of related objects arrayed together like the luxury goods in arcade shop windows. It is the look which counts. Signs float away from that which conditioned their production, and the look becomes a signified as it, too, is commodified in the trading of reputations, images and representations.

Modernist utopias and dystopias

Baudelaire emerges from his writing fraught with contradictions: critic of modern art and life, ex-aristocrat dressed as a dandy and at times as ex-convict; who identifies modernity with the crowd yet observes it from a safe distance, seeking the privacy of intimate encounters and intensity of libidinised sensations in the face of mediocrity, decay and corruption; a revolutionary who finds mass society distasteful; a stylistic innovator in his verse who is at odds with the modernisation of Paris around him. But society was also fraught with contradictions between respectability and its flouting in trade or the sexual practices of buyers and sellers of gratification, and a massification of its opportunities at the same time as greater restriction was placed on the lives of the poor. Perhaps a fitting emblem of the period is the neoclassical allegory *Le Téléphone* at the Hotel de Ville, Paris (Warner, 1985: fig. 18), stretching an idiom to breaking point in a contemporary female figure in classical-style drapes holding the receiver of a piece of undeniably modern, machine-made equipment designed to facilitate a form of distanced communication in which people speak to each other but are not face to face. The image contrasts with the use of personification made, for Marina Warner, by Baudelaire in his writing, 'to pierce us with the vivid array of his sufferings, with the grim, dreaming figure of Boredom, and the turbulence of his Sorrow and the malignancy of Beauty, his muse' (Warner, 1985: 82).

Baudelaire's collection *Les Fleurs du mal* (1857) was rejected by his publisher, and six of the poems suppressed by court order. The publisher went broke and Baudelaire went into exile in Brussels, later returning to Paris but dying in poverty. His art criticism establishes *modernité* as the cultural representation of the experience of modern life, and his poetry is taken as the beginning of an aestheticisation which rejects materiality in favour of a withdrawal from the life of the street which is key to his art criticism. He writes that 'To be a useful person has always appeared . . . particularly horrible' (Baudelaire, 1989: 27), expressing a sentiment that writers such as J. K. Huysmans and Oscar Wilde build into a new aesthetic, Huysmans's *A rebours* being a classic of the genre. But this is political, too. Baudelaire's 'disdain for public life' as T. J. Clark puts it (1973: 181) produces wasted time, an intentional wasting of time taken to the extreme of an art form by the Situationists in the 1960s, also in Paris, in an aimless drift (*dérive*) around the city – a refusal of productivity in a society based on the imperative to produce. Simon Sadler writes:

> The situationist 'drifter' was the new *flâneur* . . . Like the *flâneur*, the drifter skirted the old quarters of the city in order to experience the flip side of modernization. And situationist writing carried over some of the *flâneur*'s cavalier attitudes: page upon page passionately denounced alienation . . . but the reader was only directed toward a deeper understanding of the ghetto-dwellers' real lives with a nonchalant wave of the hand. Situationists mythologised the poor as fellow travellers on the urban margins.
>
> (Sadler, 1999: 56)

Baudelaire's withdrawal was more purposeful. In his review of the 'Salon of 1846' he called for the celebration of modern life in art. Clark sees Courbet's *Burial at Ornans* (1848–50, Paris, Louvre), a group of villagers at a funeral, its plain visual language and alignment of the figures on a horizontal representing a democratic society, as articulating this celebration – though Baudelaire did not respond to the painting exactly in this way (Clark, 1973: 181). Clark establishes the link between Baudelaire and the painter and satirist Honoré Daumier, to whom Baudelaire gave a fair copy of his poem 'Le Vin des chiffonniers' based on Daumier's *The Ragpicker* (1844?, lost – Clark, 1973: 161). I cannot do justice to Clark's reconstruction of the milieu of Baudelaire, Daumier, Delacroix and others in the 1850s, but cite this remark: 'as so often in Baudelaire's city poems, there appears a clinching metaphor which hints at a kind of travestied class relationship' (Clark, 1973: 163).

Baudelaire writes of Paris as a fantasy tinted by melancholy. He makes a voluntary identification with the ragpicker (one of the poorest categories of labourer in a typology which could be compared to that of the Chicago School's social research – see Chapter 1) against the grain of his social background but at the price of abstracting the category – the ragpicker, not identified by name. Perhaps the distancing of specific ragpickers by this literary device is loosely akin to Baudelaire's recognition of an inherent failure to find the satisfactions offered by desire. Benjamin notes, citing *Les Fleurs du mal*, 'the expectation roused by the look of the human eye is not fulfilled' (Benjamin, 1997: 149), which links *ennui* (the boredom or anxiety of repeated unsatisfaction) to commodity fetishism. The cultivation of *ennui* as a literary quality depends, of course, on the time and conditions in which to reflect on it, while commodities draw on a generalised dissatisfaction to promise delights they cannot deliver other than in the look or the idea. Often Baudelaire is read as wallowing in this, but his work has a political dimension.

Baudelaire's poems are called socialist by Jean Wallon in *La Presse de 1848* (Clark, 1973: 163). Clark asks 'Why should not *Les Limbes* (Baudelaire's title for *Les Fleurs du mal* then) have been Socialist?' (Clark, 1973: 164), noting Fourierist overtones and that the work was announced in a leftist paper and timed for publication on the anniversary of the 1848 uprising in Paris. Clark settles for ambivalence, but Donald writes:

> Somewhere in the imagination of modernity there is a third city, in addition to the city of *flânerie* and the planned technological city. This is the *republican* city, the city which since Aristotle at least has been supposed to provide the forum for debates about what constitutes the good life . . . the creation of society and the formation of self made possible only through the *civis*, may appear sometimes in surprising and profane forms . . . even if it is forever betrayed by the grimy reality.
>
> (Donald, 1999: 96)

Or it is superseded by his wish, in the third version of a preface to *Les Fleurs du mal*, 'To know nothing, to teach nothing, to will nothing, to feel nothing, to sleep and still to sleep' (Baudelaire, 1958 [1868]: xvi). In the 1880s such feelings translate into the withdrawal of Symbolism and decadence (alluding to the decay of a society, as in the late period of Rome – the republican period of which history informed the revolutionaries of 1789 via David's paintings).

Benjamin cites a text by Jules Laforgue in which Baudelaire is described as 'the first to speak of Paris from the point of view of one of her daily damned' in a world of 'lighted gas jets flickering with the wind of Prostitution, the restaurants and their air vents, the hospitals, the gambling . . . the cats, beds, stockings, drunkards, and modern perfumes' (Laforgue, 1956: 211, in Benjamin, 1999: 246). There is a fugitive aspect to Baudelaire's life: he goes to cafés, Benjamin tells us, to avoid his creditors; he has two apartments but stays with friends when the rents are due (Benjamin, 1997: 47). Situations are fleeting, like the light in Impressionist scenes of cafés and bars, the river, the boulevards teeming with carriages and pedestrians, or the value of speculative commodities.

Cultural capitals

Paris

Benjamin calls Paris the capital of the nineteenth century (1997: 155–176). The city is the centre of new movements in art, and of a cultural life also including entertainment and the society of cafés and bars. Other cities hosted cultural innovation – Barcelona had a thriving cultural and café life, and a proliferation of new architecture; and secessions took place in Vienna, Berlin and Munich in the 1890s; Futurism began in Milan in 1909, and Dada in Zurich in 1916 – but the Futurists published their first manifesto in Paris, in *Le Figaro*. This requires explanation.

Paris, Berlin, Vienna and London were imperial cities and major centres of commercial life, but they did not have equivalent cultural ambiences. The founders of the Vienna Secession met at two cafés – painters at Café Sperl and architects and designers at Café Zum Blauen Freihaus – where they fervently discussed 'everything that was new and exciting' (Vergo, 1975: 19–21). In the 1920s Germany and the Soviet Union were sites of experimental film and theatre, while the Bauhaus in Weimar (later in Dessau) united art, craft and technology in a utopian prospect of designing a better world. Yet Paris is pre-eminent in art history and the popular imagination – most North Americans who came to Europe to write, paint or study culture went there, for example. Perhaps a continuity of cultural production from the 1850s to the 1930s despite major political upheavals allowed this, together with the scale of cultural life. By comparison Vienna was an intellectual village mired in the contradictions which marked the decline of the Austro-Hungarian empire, Berlin was the node of a recent national identity defeated disastrously in 1918 and London was marked throughout by a suspicion of French ideas among the upper classes, as if tainted with immorality and revolt. Modernism, in any case, was international – and the presence in Paris of artists and writers from a spectrum of countries added a new sense of the cosmopolitan to the metropolitan experience.

Despite an arms race and looming international crisis prior to 1914, cross-border currents were characteristic of modernism. For example, Henri Matisse was profoundly affected by an exhibition of Islamic art he saw in Munich in 1910 (Jacobus, 1973: 33), before going to Morocco; and his work was exhibited in Germany, Russia and the USA as well as Paris. But if the internationalism of this period informs international modernism later, in the inter- and postwar periods, there were economic factors in the centrality of Paris as well, and later New York, as art marketplaces. Artists depend on dealers just as writers need publishers, and Paris offered a concentration of them, with the necessary contacts to collectors who were the source of livelihood for emerging artists.

That many of the works bought by collectors depicted rural, coastal or suburban scenes emphasises economic reality. To demonstrate the ambiguities of urban and rural subject-matters, however, I cite a body of work by the Impressionist painter Camille Pissarro, one of several French artists in the 1880s with anarchist sympathies. John House notes that 'Pissarro's Anarchist beliefs were firmly based on a return to the soil' (House, 1979: 16), which follows celebratory depictions of rural life by Courbet and Millet (counter to the modernisation of Paris). The boulevards were lined by apartments for the bourgeoisie, and opened with unveiling ceremonies like those of public monuments. Modernisation did not stop in 1870 with Napoleon III's exile in England, either, but extended to the opening of the Eiffel Tower in 1889 and a proliferation of statues (Michalski, 1998: 13–55) including Jules Dalou's monument for motor-car pioneer Emile Levassor (Porte Maillot, 1907) and Frédéric-Auguste Bartholdi's *Balloon of Ternes* (Place de la Porte des Ternes, 1906, destroyed 1942). Pissarro rented hotel rooms and apartments on high floors for the views of the boulevards they afforded. He would rather have painted rural scenes but there was a market for pictures of the boulevards. Richard Brettell writes:

> The economic success the urban paintings enjoyed was presumably a motivating factor in [his] decision to paint so many of them. Quite simply, they sold better and more quickly than any paintings he had ever produced, enabling the elderly Pissarros to pay off their debts, support their children and enjoy a life of unprecedented comfort.
>
> (Brettell, 1993: xxviii)

These images illustrate a gap between an intention on the part of a cultural producer and the reception of a genre in the marketplace. It appears art is as much economically as it is aesthetically determined, and that the model of an aesthetic judgement reflecting dominant taste is itself economically produced.

Case study 3.2

From David Crouch, *The Art of Allotments*

The north-east of England has a mythical status in popular allotment stories, and is often regarded as the key – if not only – place where there are, and have been, allotments. The reason for this is that the north-east is the place for the eccentric and committed pursuits of leek-growing and keeping pigeons. Some sites are crowded with so many bird lofts that they resemble Wild West shacks, often brightly decorated to guide the birds on a secure flight home. The north-east is rightly renowned for its leek growing. Plants are grown to prize-winning girth, with the use of curious potions. Some communities have competitions bearing International Trophies and competitors come from within a twenty mile range. Pigeon races can start in France and people use birds bought from Japan, but leek competitors rarely travel to the event from beyond the immediate community.

The tradition of leeks and pigeons has been an important part of the coal mining culture that extended from the north-east down to Nottinghamshire and into South Wales and Kent. It was in the north-east, however, that a unique local group of miners turned to painting and drew on their local environments for their work. Pigeon crees provided a fascinating object for study. Here the same culture represented itself through allotment cultivation and through paintings of that allotment work and its cultural identity; both rich components of local popular culture, exemplified in the Ashington Group.

(Crouch, 2003: 16–18)

Comment

Allotments are traditional sources of food for urban working-class and lower middle-class people. In Britain many were sited in the marginal land along railway lines until it was sold as a consequence of rail privatisation. But allotments are also sites of collective identity, with clear guidelines or rules as to what can be grown (vegetables, or vegetables and flowers, for instance) and what kind of structure can be built (huts of certain dimensions), and often a strong network of mutual aid in seed and crop exchange. The case upsets the demarcation between culture as the arts and cultures as a way of life in that, thinking of Raymond Williams's definitions (Chapter 1), growing vegetables is an original meaning of culture; and, to go beyond the obvious, the way in which allotment growers operate constitutes a cultural formation as distinctive, and it could be argued as aesthetic, as art or craft work. The art critic William Feaver provides a link between everyday social formations and art when he draws attention to the work of pitmen painters in the Ashington Group (cited by Crouch, above – Feaver, 1993). There have been films about allotments, too, such as *The Ballad of Ten Rod Plot* (1992) with music by the Albion Band (Crouch, 2003: 19); and newspaper articles in which allotments have taken on a bijou character – Crouch records that the *London Evening Standard* 'celebrated allotment fashion with rural-Islington chic' (ibid.). Allotments are multicultural, too, with imported as well as native species of vegetable and plant. For some they are the basis of an informal-economy business, for others a hobby, or necessary source of food. Crouch summarises, 'Most people turn their sod because they want to grow food, enjoy the "crack" and expand their imagination' (Crouch, 2003: 22). I want to put this example of self-organised production in another context. Chris Coates records that in 1895 the Clousden Hill Free Communist and Cooperative Colony took a twenty-year lease on 20 acres of land at Clousden Hill Farm in Northumberland, with support from the Independent Labour Party and the Cooperative Movement (Coates, 2001: 203, 271). Two of the activists, William Key and Frank Kapper, invited Kropotkin to be its treasurer (he declined). The Colony's statement of objects and principles includes:

> The acquisition of a common and indivisible capital for the establishment of an Agricultural Colony;
>
> The mutual assurance of its members against the evils of poverty, infirmity and old age;
>
> The attainment of a greater share of the comforts of life than the working classes now possess;
>
> The mental and moral improvement of all its members;
>
> . . .
>
> To demonstrate the productivity of land under intensive cultivation.
>
> (cited in Coates, 2001: 203–204)

In 1897 the anarchist paper *Freedom* reported that fifteen men, two women and two children were living at the Colony, and that more women were needed (Coates, 2001: 204). Internal disagreements made consensus decision-making too difficult and the Colony ceased trading in 1902 when some members moved to Whaggs Lodge Estate near Gateshead, renting land from the Northern Allotments Society.

Milieux

To defend themselves against the economic power of the market, and for companionship as strangers, artists formed loose groupings. Raymond Williams notes that many contributors to the development of modernism were migrants, and that perceptions of the artist were a product of mass media:

> The movements are the products . . . of changes in public media. These media, the technological investment which mobilized them, and the cultural forms which both directed the investment and expressed its preoccupations, arose in the new metropolitan cities, the centre of the also new imperialism, which offered themselves as transnational capitals of an art without frontiers. Paris, Vienna, Berlin, London, New York took on a new silhouette as the eponymous City of Strangers, the most appropriate locale for art made by the restlessly mobile emigré or exile.
>
> (Williams, 1989: 33–34)

Being an exile is not the same as being an outsider but sets the ground for a perception of cultural producers as outside the dominant society. In 1950s New York the modern artist or writer was mythicised as a dysfunctional individual (white, male in the received story) putting creative freedom before social status or money. For Donald Kuspit (1993) this was the modern artist's social contract.

It could be argued that the culture (way of life) of artists is neither more nor less interesting or intricate than, say, that of accountants, doctors, restaurant and bar staff, sex-industry workers or civic administrators. Paris and New York had large numbers of all these yet are represented in history and tourism as art cities more often than, say, hubs of new ideas in accountancy. In part this is a continuation of the mythicisation of the city by artists' and writers' groups themselves. In New York in the postwar years the growth of artists' groups rested on the support of a circle of dealers, critics and collectors (as in Paris in the 1870s). Dore Ashton describes the situation for artists and writers who became successful:

> They felt they were on the verge of being incorporated . . . into the mainstream of society, and they felt more isolated than ever . . . The problem of success still seemed the canker in the soul of the rebellious American artists and writers.
>
> (Ashton, 1972: 195)

She continues to describe the formation of a heterogeneous community of artists meeting regularly. But while earlier European formations tended, as Ashton explains, to meet on the basis of shared ideas, the New York artists debated from divergent positions:

When public sessions were held, everyone turned up expecting chaotic, excited, and often ridiculous exchanges among recognized antagonists, and they revelled in talk itself as a mode of self-affirmation. To answer the American vernacular greeting 'What's going on?' these New York aspirants to an American movement could only answer that something *was* going on . . . [for] the survival of the old bohemian values.

(Ashton, 1972: 196)

Those values operated in the public lectures and meeting in a tavern afterwards arranged on Friday evenings by a group of painters including Mark Rothko, Barnett Newman and Robert Motherwell; when these fora were subsumed into regular academic life the group continued to meet, joined by intellectuals including Hannah Arendt and the critic Harold Rosenberg, to form the Eighth-Street Club. Ashton remarks, 'the Club functioned, as all its members agreed, as a surrogate Parisian café' (Ashton, 1972: 198). The pattern of a new cultural class in which members from disparate backgrounds unite in pursuit of cultural aims is probably now assumed in the making of a cultural city. Cities which offered a suitable environment for this were more likely to be cultural hubs, but the growth of these milieux is outside planning and mainly outside policy, contingent more on markets for new work. In time the loose formation of patrons and artists becomes, together with intermediaries such as dealers, museum curators and critics, a cultural class among whom an informal consensus determines what is accepted as contemporary art.

Cultural capital

The main possession, as it were, of a cultural class is cultural capital – the knowledge of culture which that class articulates. It is a symbolic mechanism in which the signifiers of value are not in a given relation to the signifieds. It is also a form of capital the possession of which compensates members of a cultural class for low levels of monetary reward. Cultural capital is distinct from money capital, then, but guarded in a similar way by those who have it. Pierre Bourdieu writes in *The Field of Cultural Production* that the boundaries of fields such as theatre and philosophy are 'protected by conditions of entry that are tacitly and practically required (such as a certain cultural capital) or explicitly codified and legally guaranteed' (Bourdieu, 1993: 43). He continues that characteristics specific to the field of cultural production are the permeability of its frontiers, and diversity of the employment it offers:

It is clear . . . that the field of cultural production demands neither as much inherited economic capital as the economic field nor as much educational capital as the university . . . However, precisely because it represents one of the *indeterminate sites* in the social structure, which offer ill-defined posts,

Case study 3.3

Figure 3.2 Men's lavatory, waterfront, Hull (photo M. Miles)

Comment

I was first shown this urinal on a visit to Hull in 1992. I went there again a few years later with a camera (only to the gents – I never saw the ladies and cannot remember if there is one). The city has a trail of images of the different kinds of fish once caught by Hull fishermen and brought to its markets – mainly as plaques set in paving or brickwork – but I found the gents more interesting than this fusion of art and local history. In contrast to the obligatory niceness of public art, the gents had a quality of human warmth. It represented culture as everyday creativity rather than civic worthiness or the perpetuation of art's mythicised status as conferring universal benefit like the proverbial Lady Bountiful. I had no need to use the facilities, and only gazed at them (but they are used). Inside everything is spotless. The white china gleams, the copper pipes sparkle in the light which washes through the glass panes in the roof, and every available space supports a plant, many of them tropical. It is a smaller version of the hothouses of Kew Gardens, or a nineteenth-century winter gardens in a seaside resort (or as projected in the design of a garden city by Ebenezer Howard), loosely inflected by the same exotic dream which informed those earlier efforts to encompass the world of growing things in rain-forests and other as-if magical sites. This is, of course, a glamorised account. In one way it represents the attendant's devotion to his work, making the best of a mundane place. In another it signifies everyday creativity, or perhaps what Joseph Beuys meant when he said that everyone is an artist. Glimpses of this creativity are neither fragmentary nor fleeting in the manner, and mannerisms, of modern art and literature. The plants are cared for to prolong their existence and to maintain the equilibrium of their growth cycles. The ephemeral seems a comparative luxury here, where functionality reigns. Yet the function for which the lavatory was built in the nineteenth century and which, presumably, pays the attendant seems secondary in my experience as a visitor to what is taken by those who proudly insist that people like me should see it as a kind of exoticism. Entering the building, I am transported to a world like that of Douannier Rousseau, the French customs officer who was invited to Picasso's banquet and whose paintings of tropical plants and animals, and sometimes Parisian everyday scenes, are exhibited in the Museum of Modern Art.

> waiting to be made rather than ready made, and therefore extremely elastic and undemanding, and career paths which are themselves full of uncertainty and extremely dispersed . . . it attracts agents who differ greatly in their properties and dispositions but the most favoured of whom are sufficiently secure to be able to disdain a university career and take on the risks of an occupation which is not a 'job'.
>
> (Bourdieu, 1993: 43)

Artworlds are characterised by mutabilities and fluctuations in the currency of particular kinds of art, and the universality of cultural value is mediated by cultural institutions whose criteria are tautological. But cultural producers cannot dispense with the validation of institutional structures, and their distinctions are the more important as members of individualistic fields seek esteem.

This does not prevent (and may allow) alliances with, or sympathies for, other dominated groups. As Bourdieu writes, 'The cultural producers, who occupy the economically dominated and symbolically dominant position within the field . . . tend to feel solidarity with the occupants of the economically and culturally dominated positions within the field of class relations' (Bourdieu, 1993: 44). The role of intellectuals varies within national cultures but Bourdieu's argument that cultural producers 'are able to use the power conferred on them, especially in periods of crisis, by their capacity to put forward a critical definition of the social world . . . and subvert order prevailing in the field of power' (Bourdieu, 1993: 44) implies a definition of cultural groups similar to that of an intelligentsia. In the twentieth century cultural institutions have to an extent taken back the role of supporting cultural formations, after a series of departures beginning with the Secessions of the 1890s or the artist-run exhibitions of Paris in the 1880s. A characteristic of modernism is the development of new modes of sociation, and I end by looking at the artists' banquet as an example of this.

The banquet

The banquet became an important social form for the bourgeoisie of Paris in the late nineteenth century. Sennett sees a new form of public–private consumption:

> A uniform dinner was served, after which two or three people would make speeches, read from their own or other people's books, or otherwise entertain the crowd. The banquet ended what the coffeehouse two centuries before began. It was the end of speech as an interaction, the end of it as free, easy, and yet elaborately contrived. The massive banquet was the emblem of a society which clung to the public realm as an important realm of personal experience, but had emptied it of meaning in terms of social relations.
>
> (Sennett, 1992: 217–218)

The banquet arose as a means to class distinction for an aspiring bourgeoisie seeking establishment in a country once again a republic after 1870, but with a nostalgia for aristocracy matching a monarchist current in political life.

If social meaning was drained from the banquet, however, it was reintroduced in an appropriation of the form by avant-garde groups in the 1900s. In *The Banquet Years*, an account of French avant-gardes between 1885 and 1914, Roger Shattuck sees the formal banquet as where 'the cultural capital of the world . . . celebrated its vitality over a long table laden with food and wine', and a chance for 'pompous display, frivolity, hypocrisy, cultivated taste, and relaxed morals' (Shattuck, 1969: 3). There were less formal banquets, too. In 1885 the Chat Noir cabaret (which published Verlaine and Mallarmé in its magazine) arranged a banquet to mark its move to new premises. The guests were writers, deputies, mayors and the academic painter Bouguereau. It was a meeting of avant-garde and civic establishments. A banquet of a more avant-garde milieu was organised by Picasso in his studio, the *bateau lavoir* – 'a huge barnlike room with thick beams and African masks on the walls' (Shattuck, 1969: 67) – to honour the painter Henri Rousseau more than a decade later. Plates and glasses were borrowed, chairs hired, food ordered and thirty guests invited including the poet Guillaume Apollinaire. The banquet began earlier in a bar, from which the guests emerged some time later. Shattuck records a 'gargantuan supply of wine':

> Finally three knocks resounded through the room. The door opened to reveal Rousseau wearing his soft felt artist's cap, carrying his cane in one hand and his violin in the other . . . After struggling more than two decades for recognition and fellowship with other artists, Rousseau looked through the door . . . and saw Picasso's studio decked out as it would never be again and filled with artists and writers who had come to honor him.
>
> (Shattuck, 1969: 68)

The drinking, songs and impromptu entertainments – including an impression of delirium tremens, produced by chewing soap, by the poets André Salmon and Maurice Cremnitz for the benefit of visiting American Gertrude Stein (Shattuck, 1969: 66–69) – lasted until dawn.

This shows the banquet as a site of that formation of a milieu for which the cafés of cities such as Paris, Berlin and Vienna were other sites. Paul Verlaine and friends, for instance, held a banquet above a café on the corner of rue Cujas and boulevard Saint-Michel on 11 October 1878 to found the Hydropaths, a group of three hundred moving on to the Café de l'Avenir in Place Saint-Michel (Stephan, 1974: 43). The magazines such groups produced never lasted long, and café society, too, was mobile – yet both underpinned modernism as the cultural production of milieux, reflecting its fleetingness.

Summary

The above gives me no exact answer as to why Paris was the cultural hub of modern art until the 1930s, and New York from the 1940s to the 1960s. The various, at times disconnected, factors I cite suggest that new art is produced in new social formations which define themselves through it. This may seem circular but my underlying argument is that such formations happen only in large cities which support a critical mass of producers, intermediaries and consumers but in which also there is a sense that to be present allows participation in something new. None of this is possible in a village, and although artists have sometimes retreated to rural life they have brought the pictures to a city to show and sell them. But perhaps this model of cultural production and reception is over, as artists increasingly turn to cyberspace to find global publics for their increasingly immaterial production; and as cities in Asia, Australasia and Latin America host art fairs and international exhibitions or other cultural events as important as any in Europe or North America.

4 Culture re-producing cities

Learning objectives

- To gain insights into the re-presentation of cities as cultural sites
- To investigate the concept of a symbolic economy
- To see how symbolic economies operate in specific cases

Introduction

In Chapter 3 I examined the production of cultural movements. In this chapter I move from cultural production to cultural consumption and its role in the reproduction of a city's image. I begin by reconsidering the concept of a symbolic economy, comparing the positions of Sharon Zukin (1995; 1996a; 1996b) and John Allen (2002). Then I investigate Barcelona's transformation from a Mediterranean port to a world city in the 1990s as exemplar of a city's rebranding through cultural means. In case studies I cite Joan Didion's essay on San Francisco counter-culture in 1967, the year of the Summer of Love, and Kim Dovey's account of the gambling-led redevelopment of Melbourne's waterfront, which can be compared to culturally led development. I end with an image of a wall painting commemorating a house-owner's pilgrimage to Mecca, in a village in the far south of Egypt. Although outside the urban I take this to demonstrate cultural frameworks and unities other than those of Western city marketing.

Cultural economies

Cultural nodes

The marketing of place depends on qualities of vibrancy or vitality, while the prominence of cities as cultural hubs follows patterns of cultural consumption in cultural tourism, and attraction of aspirant arts professionals as producers and consumers of a city's image. Sharon Zukin sees artists as 'a cultural means of framing space . . . [who] confirm the city's claim of continued cultural hegemony, in contrast to the suburbs and exurbs' (Zukin, 1995: 23). She adds that artists' presence 'puts a neighbourhood on the road to gentrification' (Zukin, 1995: 23, citing Zukin, 1989; Deutsche, 1991) as happened in SoHo and the Lower East Side in New York, and notes increasing politicisation and self-conscious defence of their interests among artists: 'Often they have been co-opted into property redevelopment projects as beneficiaries, both developers of an aesthetic mode of producing space . . . and investors in a symbolic economy' (Zukin, 1995: 23). That investment may be intentional as promotion becomes integral to contemporary art practice, or less so when artists are opportunists finding cheap spaces in which to live and work – such as the short-life housing colonised by artists' organisations in London from the 1970s to 1990s and seen in retrospect as prelude to gentrification.

Zukin remarks that 'every well-designed downtown has . . . a nearby artists' quarter' (Zukin, 1995: 22). The aim of developers, businesses and property owners is to increase economic activity by trading on the attractions of cultural value. But for artists the recoding of deindustrialised or marginal spaces as production sites is part of an aim to take control over their own futures and means of production and distribution. For Raymond Williams modernism is characterised by self-naming artists' groups ranging from the establishment of 'normative groups which sought to protect their practices within the growing dominance of the art market and against the indifference of the formal academies' (Williams, 1989: 50) to 'more radically innovative groupings, seeking to provide their own facilities of production, distribution and publicity' (Williams, 1989: 50–51) and 'fully oppositional formations, determined not only to produce their own work but to attack its enemies in the cultural establishments and . . . the whole social order' (Williams, 1989: 51).

Williams's scenario indicates a situation in which an intention to reproduce space as a means to take over the means of cultural production and distribution is fused today with development agendas which artists cannot control. The presence of artists does not as such cause gentrification, but is a signal of it leading to a peripheralisation of artists when property values and rents rise as a consequence.

Zukin sees artists as co-opted to development agendas by business elites in finance, insurance and property speculation – 'generally great patrons of both art museums and public art, as if to emphasize their prominence in the city's symbolic economy' (Zukin, 1995: 23) – while Pierre Bourdieu views artists as having a 'particularly lucid perspective on the threats that the new economic order represents to the autonomy of the intellectual "creators"' (Bourdieu, in Bourdieu and Haacke, 1995: 16). He continues that the rise of private-sector patronage offers corporations an aura of 'disinterested generosity', and that 'There is . . . an extremely perverse mechanism which operates in such a way that we contribute to our own mystification' (Bourdieu and Haacke, 1995: 16–17). This contributes to a symbolic economy when new bohemians (Wilson, 2000; Lloyd, 2006) occupy a culturally re-coded district in which a city's edges – as between privation and aestheticisation – may be part of the attraction.

Symbolic economies

To gain a place on the list of world-class cities requires both symbolic capital and money capital. Symbolic capital accrues to blue-chip art institutions, signature architecture, loft living spaces, fashionable shops and restaurants and designer bars. Like the status of art in its market, a city's status depends on value judgements among cultural and marketing intermediaries. Zukin explains that cities have always had symbolic economies in that elites have manipulated symbolic languages to mirror a predetermined image of a city. She separates the symbolic economy of dominant representations of a city from the political economy of material conditions of groups in society. Zukin defines a symbolic economy as 'The look and feel of cities [which] reflect decisions about what – and who – should be visible and what should not, concepts of order and disorder, and on uses of aesthetic power' (Zukin, 1995: 7).This immaterial production can be compared to Michel Foucault's account of the exclusion of vagrant and insane people from the streets of Paris and their confinement in the Hôpital Général in 1656, in *Madness and Civilization* (1967). The non-productive are removed from the visible city so that its dominant image denotes productivity, while 'philanthropy prefers to recognize the signs of a benevolence toward sickness where there is only a condemnation of idleness' (Foucault, 1967: 46). Zukin writes of business elites whose 'philanthropy, civic pride, and desire to establish their identity as a patrician class' lead them to invest in museums, parks, and architecture to signify a world-class city (Zukin, 1995: 7–8), while culture 'offers a coded means of discrimination, an undertone to the dominant discourse of democratisation' (Zukin, 1995: 9).

Symbolic economies combine two parallel production systems: 'the *production of space*, with its synergy of capital investment and cultural meanings, and the

Case study 4.1

From Joan Didion, *Slouching Towards Bethlehem*

At the time I was in San Francisco the political potential of what was then called the movement was just becoming clear. It had always been clear to the revolutionary core of the Diggers . . . and it was clear to many of the straight doctors and priests and sociologists who had the occasion to work in the District . . . the boys in the rock groups saw it, because they were often where it was happening. 'In the Park there are always twenty or thirty people below the stand,' one of the Dead complained to me. 'Ready to take the crowd on some militant trip.'

But the peculiar beauty of this political potential, as far as the activists were concerned, was that it remained not clear at all to most of the inhabitants of the District, perhaps because the few seventeen-year-olds who are political realists tend not to adopt romantic idealism as a life style. Nor was it clear to the press, which continued to report 'the hippie phenomenon' as an extended panty raid; artistic avant-garde . . . or a thoughtful protest . . . against the culture which had produced Saran-Wrap and the Vietnam War . . .

Because the signals the press was getting were immaculate of political possibilities, the tensions of the District went unremarked upon . . . The observers believed roughly what the children had told them: that they were a generation dropped out of political action, beyond power games, that the New Left was just another ego trip . . . Of course the activists . . . had long ago grasped the reality which still eluded the press: we were seeing something important. We were seeing the desperate attempt of a handful of practically unequipped children to create a community in a social vacuum. Once we had seen these children we could no longer overlook the vacuum, no longer pretend that the society's atomization could be reversed.

(Didion, 2001 [1967]: 103–104)

Comment

The District to which Didion refers is Haight Ashbury in San Francisco, and the rock group is the Grateful Dead. Her account (written in 1967) was based on living near Haight Street for several months in the year of the Summer of Love. Didion uses the term *children*, and opens her account by saying that adolescents began drifting from city to city, and that 'People were missing. Children were missing. Parents were missing . . .

San Francisco was where the missing children were gathering and calling themselves "hippies"' (Didion, 2001: 72), attracted by an open lifestyle in which music, free love and soft drugs were elements; but not the only elements, as Didion makes clear. Within the melting pot of an informal and partly self-organising society were political currents, and tactics such as mime and street theatre through which to provoke reactions. Didion notes that the Diggers emerged from the Mime Troopers during the Hunter's Point disturbances of 1966, 'when it seemed a good idea to give away food and do puppet shows in the streets making fun of the National Guard' (Didion, 2001: 106). Much of the daily life of Haight Ashbury was confused and politically inarticulate, and the 200,000 drop-outs provided a market for suppliers of various substances. Yet the external perception, particularly in the mass media, was of a youth rebellion concerned with lifestyles and music rather than power. A city known for hippies – a few of whom were given free coffee by cafés wishing to attract visitors to the human zoo – was not a city in revolt against the values and culture of a nation-state with a poor human rights record for people of colour, currently engaged in a brutal war in Indo-China. If culture is the vocabulary within which we act (Belsey, 2005 – see Chapter 2), the culture from which the subculture of San Francisco withdrew its consent was that of a nation-state of relatively recent and dubious origin in the 1860s, which privileged private enterprise over state regulation and seemed increasingly contradictory. The Summer of Love did not achieve political change, but it is important in retrospect not to accept the rosy image of a city giving liberalism a cool and smoky turn to the sound of rock music. If Paris, New York and Barcelona gain cultural hegemony through mainstream cultural movements and institutions, San Francisco trades on a past as a counter-culture hub but depoliticises this to assimilate subversion in place marketing. A similar assimilation has taken place in rock music generally (Seiler, 2000). Yet the ethos of the counter-culture has not died, and continues (if only in the imaginations of people such as myself, an art student in the late 1960s) to evoke a latent utopianism, a thought that a world indisputably different and kinder, a world of love not war, is possible. The point, which leads me to include this case, is that its realisation will be political. But then, the realisation of a city's global aspirations, say as a cultural capital, is also political. It is merely that the political agendas and allegiances differ.

production of symbols, which constructs both as currency of commercial exchange and a language of social identity' (Zukin, 1995: 23–24). A symbolic economy becomes important as traditional economic sectors based in material production decline and cultural industries dealing in immaterial production (in the media and public relations as well as the arts) grow. Zukin worries that the rise of a symbolic economy displaces demands for material gain so that:

> Styles that develop on the streets are cycled through mass media, especially fashion and 'urban music' magazines . . . where divorced from their social context, they become images of cool. On urban billboards advertising designer perfumes or jeans, they are recycled to the streets, where they become a provocation . . . The cacophony of demands for justice is translated into a coherent demand for jeans.
>
> (Zukin, 1995: 9)

Coherence is provided by the market when symbolic values attach to particular goods and services denoting consumers' affiliation to a social group. In *Soft Cities* Jonathan Raban writes of a vogue for white-painted Moroccan bird cages in London in the 1970s (Raban, 1974: 95) among people who did not keep birds but saw the empty cages as accessories to interior design. Such purchases are validated through the perceptions of a peer group (Lury, 1996: 14, citing Douglas and Isherwood, 1979: 75) in a way resembling the validation of contemporary art by its institutions. Mike Featherstone (also citing Douglas and Isherwood, 1979) reads such consumption as marking social boundaries, adding, 'the mastery of the cultural person entails a seemingly "natural" mastery . . . of how to use and consume appropriately with natural ease in every situation' (Featherstone, 1995: 21). Hence those admitted to symbolic consumption are at risk of failure: 'Their cultural practices are always in danger of being dismissed as vulgar . . . by the established upper class . . . and those "rich in cultural goods" – the intellectuals and artists' (Featherstone, 1995: 23). Featherstone continues that 'the specialists in symbolic production will seek to increase the autonomy of the cultural sphere and to restrict supply and access to such goods, in effect creating and preserving an enclosure of high culture' (Featherstone, 1995: 23). He concludes that the demand for cultural goods 'must be understood within a social framework' that emphasises its relational character (Featherstone, 1995: 24).

Zukin sees the market as leading cultural consumption. Writing of business services, cultural industries and real estate development in New York, Zukin says that 'These activities give a much larger creative role to the organization of consumption' while 'consumption spaces . . . play a more important role in people's lives than was thought before' (Zukin, 1996b: 240). In *The Cultures of Cities* she sees real estate and the culture industries (as she calls them – see Chapter 5) as instrumental in strategies for development (Zukin, 1995: 118–133). Yet cultural entrepreneurs do not require a pool of cultural producers. Cultural consumption seldom takes place in the same site as cultural production, and Diane Dodd remarks of Barcelona, 'the creative or production side of the arts has been neglected' (Dodd, 1999: 61). A symbolic economy, then, separates production from consumption, the symbol from the conditions of its production, so that the image of a city floats over the city's streets and the life taking place therein, yet perhaps not altogether away.

The symbol requires the city as ballast but informs decisions, as Zukin says, on the production of space in a migration of culture into mainstream economic operations. But is Zukin's emphasis on culture appropriate? John Allen argues that Zukin sidesteps 'the cognitive realm of patented reason and technology' by endowing some but not other economic sectors with symbolic significance; and that 'it is simply misleading to equate the symbolic with the cultural' (Allen, 2002: 39). Allen's point is not that aesthetic goods do not figure in global competition among cities but that all economies are symbolic, including those of financial services, accounting, insurance and information. He continues:

> Equally, in the much vaunted technologically innovative sectors of tele-communications, engineering and computing, economic knowledge rests upon a combination of symbolic functions and uses which shares much with the symbolic schemas that routinely shape the practices of the recognized 'cultural' industries.
>
> (Allen, 2002: 40)

All economic activity, then, has abstract and expressive registers. A sector's distinctiveness for Allen is in how it combines these with cognate rationality in a situation in which all economies evince dexterity of symbolic knowledges. Allen cites Scott Lash and John Urry (1994) to the effect that the symbolic content of goods is privileged in new areas of consumption over their material content; and that the production of signs has two parallel modes, one cognitive in flows of information, the other expressive in areas of representation. He comments, 'Significantly, Lash and Urry do not restrict symbolic work to a particular sector of the economy' (Allen, 2002: 41), recognising design as integral to manufacturing as well as to fashion or consumer services. But Allen criticises Lash and Urry for separating cognitive from aesthetic signs on the basis that the manipulation of signs entails cognitive reflexivity which uses the understanding that analytic principles are open to question and renegotiation, and aesthetic reflexivity which uses a hermeneutic approach whereby 'subjects . . . are actively involved in the construction of their own identities through their engagement with lifestyle and consumer choices' (Allen, 2002: 41).

The cognitive deals in judgement, aesthetics more in taste. Allen reads this as tending to a potential evacuation of non-aesthetic economic sectors of what he regards as a generic possibility of fusion between cognitive and expressive modes of understanding economic activity – 'the creative play of symbolic work is not limited to a particular sector of the economy' (Allen, 2002: 43). This appears a valid argument, easy to understand when applied to the financial futures markets in which value is conjured in projections. But it applies, too, in less obvious areas such as insurance, where the cost of future risks can be no more than a projection in a complex web of mutually inflecting factors. This does not mean that activities such as city marketing, which are themselves creative, are not distinctive in trading not only on projections and the manufacture of reputations but also on real cultural histories. The histories are mediated and packaged, the qualities mythicised, of course, as signs of a city image floating freely above the city's built and social actualities, but at some point the image needs a basis in actuality to have credibility. This explains why models of city marketing based on symbolic economies are not transferable. So financial markets trade on a combination of intrinsic and extrinsic values. A symbolic economy trades on image, an extrinsic perception abstracted from the intrinsic conditions in which the image is made. Zukin notes the links between developers and cultural institutions, and that such institutions are themselves developers (Zukin, 1995: 109–152). She writes that the dependence of cultural institutions on private money 'produces a creative tension between high culture and speculative, partly unregulated, "wild" commerce' (Zukin, 1995: 119). But other milieux have a creative side and culture of their own.

Barcelona

World city

Barcelona is 'hailed as the most successful global model for post-industrial urban regeneration based on its urban design' (Degen, 2004: 131). The city promotes itself through an architectural heritage which spans the medieval and the modernist (here an equivalent of Art Nouveau); museums for Picasso, Miró and Catalan as well as contemporary art; performing arts venues and popular music; winding alleys; a revitalised waterfront; and consumption of food, drink, fashion and contemporary art in a cultural quarter. There is a World Trade Centre designed by I. M. Pei – a medium-rise, circular building with a hotel, cruise ship terminal and financial industries spaces on the redeveloped waterfront – to assert a claim to world-city status.

The Catalan day of mourning is 11 September, the day the city fell to the Bourbons in 1714. Dirges are sung in a newly created public space near the church of Santa Maria del Mar, one of many created in the city's preparations for the Olympic Games in 1992. Such histories are potent in popular conceptions of its identity. For some of the city's older inhabitants allusions to the 1930s are ingrained, such as the iconic image of three tall chimneys of a power station at Sant Adria de Besós, one of the last sites of anti-fascist resistance; or the memorial to citizens shot by the fascists in the cemetery on the far side of the hill of Montjuic. Barcelona's history as a Socialist city after 1975, too, can be traced in street names – Jardins de Rosa Luxemburg and Plaça Salvador Allende.

The Olympics

Barcelona hosted the 1992 Olympic Games as a way to re-present itself after its oppression from the fall of the Republic to Franco's death in 1975. Its transformation from Mediterranean port to world city raises a number of issues relevant to this chapter: how Catalan culture has been utilised in marketing the city globally; how flagship cultural institutions have figured in the process; how culturally led redevelopment integrated the cultures of its publics with the cultural attractions of its tourism; and how public benefit is reconciled with the requirements of global competition for tourism and investment. In Barcelona's preparations for the 1992 Olympics more than a hundred new public spaces were created in residential neighbourhoods throughout the city (Ghirardo, 1996: 200). The city authorities' intention was to spread the benefits of urban design widely, renewing the hitherto declining waterfront and opening the beaches to public use – in a tradition of enlightened city planning which began with the plan for the city's northern extension (*Eixample*) by Idelfons Cerdà in 1859. Cerdà produced treatises on urbanisation, using the term to mean a benign ordering of urban space. Today citizens of all social classes use the beaches. But some sections of the waterfront have a more affluent ambience, with globalised, American-language signs such as Beach Club, and brands of fast food and fizzy drinks; and in the furthest section of the waterfront from the old port, redeveloped for the Universal Forum of Cultures in 2004, is a round, space-age McDonald's denoting a market-led attitude which seems to have replaced a public-benefit ethos more recently.

The Olympics were not the first international spectacle hosted by the city, which held a Universal Exposition in 1888 (Meller, 2001; Degen, 2004) and an International Exhibition in 1929 for which Mies van der Rohe designed the German pavilion, now reconstructed at the base of Montjuic. Sporting facilities were developed for the 1982 football World Cup. But the Olympics were the most prominent opportunity to re-present the city internationally after the end of the fascist period. The new public spaces were generously landscaped, and many were sites for public art – some commissioned from Catalan artists, one major work from the Basque artist Eduardo Chilida (Parc dela Creueta del Coli, 1986), and some sourced internationally: Richard Serra (Palmera de Sant Martí, 1985); Roy Lichtenstein (Moll de Bosch, 1987); Beverly Pepper (Parc de l'Estació del Nord, 1992); and Claes Oldenburg and Coosje van Bruggen (Parc de la Vall d'Hebron, 1992). Serra designed a characteristic wall of rusting steel slicing across the plaza. Oldenburg and van Bruggen designed a large, painted steel sculpture of bookmatches (fig. 4.1), sited near new apartments to house Olympic athletes in the north of the city. It is red and yellow – the colours of the Catalan and Spanish flags – and perhaps its single upright, flaming match amid the bent

Case study 4.2

From Kim Dovey, *Framing Places*

The new Crown Casino [in Melbourne] opened in 1997 on a site comprising 7 hectares of formerly public land with half a kilometre of prime river frontage facing the city. The design of the building was also the subject of a secrecy agreement . . . The 7 hectare site quickly grew to encompass surrounding blocks with additional hotels, theatres, car parks and shopping malls . . . The word 'casino' disappeared from the advertising of what is now 'a world of entertainment' with theatre, conference, cinema, restaurant and retail facilities.

. . .

The most significant aspects of the design were not controlled by the architects. The 5 hectares of gaming floor are almost entirely internalized and subdivided internally along class lines – from the local working-class poker machines at one end to the exclusive 'high roller' rooms several hundred metres away to the east. The interior design of the building has been undertaken with attention to the principles of *feng shui* and a 'courtesy bus' trawls regularly through the local Chinese district for customers. All entrances from car parks are filtered through the controlling syntax of a shopping mall which also mediates the relation of casino to riverfront. The eruptions of joy marketed on the logo now erupt in an hourly spectacle of huge gas fireballs from a series of towers lining the riverfront. This is essentially advertising masquerading as a fireworks display; public land turned into the world's longest billboard. This is a dramatic spectacle of persistent carnival – the illusion of exciting urban dynamism coupled with the archetypal energy of fire. The television advertising super-imposes carnivalesque images of a giant masquerade with those of the building and its explosions of fire. As a spectacle it constructs a space for 'play' – theatre, indulgence, exuberance, leisure and liberty – the antithesis of work, constraint, poverty and addiction.

This literal form of casino capitalism has now addicted both the state and many of its citizens to gambling. Evidence of gambling related poverty and suicide mounts as consumption shifts away from local communities and retail spending. The state becomes increasingly dependent on gambling taxes.

(Dovey, 1999: 163–164)

Comment

Dovey presents blatant commercialism in the reframing of urban space, allied to the use of spectacle in the gas flares which light up the waterfront. Steven Miles notes gambling interests as linked to the tourism industry in Australia, while in the USA a disdain for the corruptions of gambling is set aside in view of the economic gains which casinos bring to a city (notably Las Vegas): 'American casinos have been commercially successful and the symbolic value of casinos to the image of cities as exciting and progressive cannot be underestimated' (Miles and Miles, 2004: 110). Miles continues that the economic impacts of the gambling industry are persuasive for politicians while local communities and their spaces are undermined when 'casinos construct their own mono-communities in the form of a one-stop consuming experience on tap' (Miles and Miles, 2004: 111). Melbourne's casino district exemplifies this in an architecture which is visually, politically and socially isolated from the city. Mark Gottdiener argues that in Las Vegas a style of built environment tailored to consumption in which gambling is a key feature is a rejection of the modernist ethos of form following function in favour of a postmodern 'victory of symbols-in-space' denoting 'unadulterated commercialism' (Gottdiener, 2000: 277). He observes a lavish vocabulary of allusions (or illusions) in the architecture of Las Vegas, from Treasure Island to the French Riviera and Egyptian-style pyramids which, he argues, puts Las Vegas in context not only of the global gambling industry but also of tourism:

> a multidimensional experience of seducing pleasures – money, sex, food, gambling, and nightlife. Las Vegas constitutes a specialized space; it is one of several global 'pleasure grounds,' such as Monte Carlo . . . and the French Riviera . . . the Greek Islands, Rio . . ., Disney World, Marienbad, or the Taj Mahal.
>
> (Gottdiener, 2000: 278)

Melbourne's aspiration, or rather that of some of its entrepreneurs and local politicians, echoes this – competing with more glamorous destinations which it apes in spectacular eruptions of flame which mask the iniquities of losing money in rooms devoid of daylight, as set apart from the street as the white-walled spaces of a modern art museum. The consumption of losing money is validated for the consumer by the fantasy of winning – free money (as a Patti Smith song says), a utopian dream transposed to an arena in which it becomes a parody. Behind the allure is the Land of Cockaigne. I end with a quote from Steven Miles in a book we co-authored:

> The amazing thing about Las Vegas is the way in which it appears to make the entire world immediately available to the consumer. In recent years, new resorts have included *Venice* and *Paris* which re-create the urban structure in a microcosmic form in order to accentuate the consuming experience.
>
> (Miles and Miles, 2004: 117)

Figure 4.1 *Claes Oldenburg and Coosje van Bruggen,* Bookmatches, *1992 (photo M. Miles)*

Figure 4.2 *Josep Sert, Spanish pavilion for the 1937 Paris World Exposition, reconstructed 1992 (photo M. Miles)*

and fallen symbolises Catalan resistance in the 1930s, if it can carry the gravity of this content. The city's political history is referenced more directly in Sert's pavilion (fig. 4.2) sited across the road.

The Olympics began a process of renewal aimed at raising the city's profile in external perception and improving the built environment for its inhabitants. This is inseparable from the aesthetic and economic transformation of the city, as from the reassertion of Catalan nationalism within the Spanish state with its restored

Bourbon monarchy, moving to greater powers of self-government in 2006. Monica Degen comments that under Franco public investment in the city 'had been deliberately neglected as a way of punishing a city that had been the bastion of the republican and anarchist movements' (Degen, 2004: 132). The Catalan language was also proscribed. After 1975 sex magazines appeared on the city's news stands, drugs were used in public squares, art student groups flocked increasingly to the Picasso Museum and the Miró Foundation, and the city's cheap bars, but there was also a determination to reassert the city's and with it Catalunya's identity, as in putting street names in Catalan.

With liberation came a strategy for public-sector-led redevelopment in a renewal of the city's urbanisation. Urban design was 'consciously taken up by the socialist city council (elected in 1978) . . . as a drive to reverse the disastrous policies of the previous regime' (Degen, 2004: 133). Design was integral to the new socio-cultural programme directed by a team of architects including Oriol Bohigas, designers, planners and engineers, led by the city's mayor, Pasqual Maragall (elected in 2003 as President of Catalunya). My first thought, in several visits to Barcelona between 1999 and 2003, was that this period of the city's renewal echoed that which followed the demolition of the city's walls and ending of its status as a fortified city (by which no building could be carried out beyond the walls), and building of the northern extension. But in conversations with academics and city planners I realised that because Cerdà had been supported by the court in Madrid rather than by local elites – eventually he left the city after years of frustrated efforts to realise the progressive order of his plan – he was still seen as an outsider. From my own perspective, in contrast, there appears a continuity from his 1859 plan to the redevelopment of the early 1990s. For that reason I make a brief detour into Cerdà's plan to question the purpose of reform.

Urban reformism

Cerdà's plan adopted a regular grid as a mark of rationality, and, although it was subsequently modified through speculative development, the infrastructure of roads and services remains largely as he envisaged it (Soria y Puig, 1999; Gimeno, 2001). Cerdà saw previous conditions within the city's walls as unhealthy and unjust. There were outbreaks of cholera, and strikes, in the 1850s, after which Cerdà, trained as an engineer, was commissioned to write a report on the city's housing conditions. His plan is conditioned by his findings, and is enlightened in providing decent housing for artisans and the bourgeoisie, public parks, street lighting and seating, pedestrian pavements taking 50 per cent of street width, and gardens within each urban block. The regular proportion of the grid was explicitly rational and democratic: 'By distributing with total equity and perfect justice the benefits of "vitality" and of building among all streets and all of the street blocks which bound them, the square grid system has the inestimable advantage of not creating odious artificial preferences' (Cerdà, 1861: 691, in Soria y Puig, 1999: 128). Like the Baixa district of Lisbon in which the emerging commercial class could live as well as trade (Maxwell, 2002), *Eixample* represents in plans and elevations the reformist ethos of that class. Cerdà aimed to provide all citizens with equal access to fresh air and green space as sources of what he terms vitality. Zoning was mixed use – apartments over shops and factories integrated with green and residential areas, linked by a subterranean rail network for goods and passengers. His plan seems progressive today but it is important to realise that his intention was to provide improved living conditions as a way to produce better citizens and lessen the prospect of unrest. The street grid with its regular proportions and octagonal intersections is one side of a rationality of which civil equilibrium is another, within the ideology of liberal reform.

The city's renewal in the 1990s echoes the public-benefit and policing aspects of Cerdà's plan. The new public spaces of 1992 provide outdoor social spaces for inhabitants, and the redevelopment of El Raval as a cultural quarter brings light into dark alleys as a way to discourage street crime and prostitution. As a local councillor said, public space provides permeability, 'the facility of penetration by the exterior' (cited in Degen, 2002: 27–28).

Catalan cultural infrastructure and cultural tourism

A major programme of new cultural spaces extended beyond the Olympics to include a National Theatre at Plaça de les Glóries (Ricard Bofill, 1997), a National Auditorium (Rafael Moneo Vallés, 1999), a National Theatre Institute (Ramon Sanabria and Lluis Comerón, 2000) and refurbishment of the Catalan Museum on Montjuic. The lead was taken by the public sector in investing in the new waterfront around Port Vell, leasing redeveloped sites to the private sector. The growth of a Catalan cultural infrastructure was, clearly, an aspect of national identity formation. But it followed, too, a policy decision in 1993 to encourage cultural tourism, formalised in the establishment of Turisme de Barcelona as a joint venture by the City Council, the Chamber of Commerce, Industry and Shipping and the Foundation for the International Promotion of Barcelona (Dodd, 1999: 54). The dual aim was to widen cultural access for residents and attract cultural tourists to the city. Mass-market tourism on the beaches of the Costa del Sol offered little to local economies when tourists bought holidays from global operators, while business executives, conference goers and middle-class consumers of city breaks, seeking to be travellers rather than tourists, are prepared to search out a city's spaces of cultural life and what might be called authentic streets and bars.

Degen writes, citing John Urry (1995), that 'Barcelona-as-concept, its mixture of avant-garde design, stylish architecture, and Mediterranean lifestyle (symbolized in open-air terraces, lively street-life, markets, and so on), can only be experienced by being *in situ*, by taking part in the everyday life of the city' (Degen, 2004: 137). And Diane Dodd that 'In striving to attract up-market middle-class tourists, the city is intent on promoting the arts and its contemporary image' (Dodd, 1999: 50). If cultural tourists seek the real Barcelona this is enhanced when the signs and information for cultural spaces are only in Catalan. Dodd argues that 'even if the tourists do not use all the attractions on offer, the knowledge of their presence . . . will encourage a return visit' while 'business or culturally educated tourists' will look for them as authentic sites (Dodd, 1999: 57–58). Barcelona's policy contrasts with Bilbao's expensive import of a Guggenheim and emphasis on international art to bypass the aspirations of regional (Basque) cultural groups while gaining the support of elites attuned to international commerce (Gonzalez, 1993). The integration of Catalan and tourist agendas in Barcelona is cost-effective. The presence of foreigners in cultural spaces does not detract from their appeal for residents while the dual agenda assists in gaining inward investment in culture (assisted by savings banks which invest profits locally in cultural institutions and art collections). Degen remarks that 'Barcelona's café culture or nightlife . . . is enjoyed as much by locals as by outsiders' (Degen, 2004: 137).

Yet there are tensions, and a rise in street crime as tourists bring affluence to a city which has pockets of deprivation still. Barcelona's young professionals now avoid the waterfront area of Maremagnum where bars, restaurants and designer shops have been replaced by fast-food outlets. Disaffected youth and immigrants spend time there, watched by security guards who can be brutal (Degen, 2004: 138) – not the authenticity cultural tourists seek. They were happier in the alleys of the old city, finding bars which had changed little since Franco's time and rubbing shoulders with artists, prostitutes, drug dealers, students and members of an ethnically diverse population.

A cultural quarter

El Raval was the red-light district and is now a cultural quarter. Its centre is the new Rambla del Raval (Pere Cabrera Massanés and Jaume Artigues Vidal, 1998–2000), 317 metres long with palm trees and rusted steel street lighting like Serra's sculptures. It vanishes at each end into old, narrow streets but fills the void created by demolition of five blocks of apartments after compulsory purchase by the city. Photographs of the neighbourhood as it was are exhibited now in the Museo de l'Eròtica (on the old Ramblas). Near the new Ramblas is the Museum of Contemporary Art (MACBA) designed by Richard Meier in a void created by

Case study 4.3

Figure 4.3 *Dosh, Egypt,* hajj *painting, 2004 (photo M. Miles)*

Comment

Dosh is a village of mud-brick houses, one of the southernmost settlements in the Kharga Oasis at the southern tip of the New Valley to which the Egyptian government hopes to relocate people as a way to lessen the strains of rapid expansion in the population of Cairo (*el-Kahira*, the city). I went there in 2004 while researching Hassan Fathy's mud-brick architecture (see case study 1.3). I include this image not because the buildings in Dosh are of a kind on which Fathy drew in general terms, however, but because it shows, in a narrow street which simply marks the space between properties, a wall painting commemorating the house owner's pilgrimage to Mecca (*hajj*), in a vernacular tradition possibly going back to the nineteenth century when items such as china plates were inserted in walls for decorative effect (Miles, 2000: 79–104). It seems to me that such traditions can be posed as alternative symbolic representations to those of world city and global city economies, and are no less elements in a coherent vocabulary of space making. But in this case the vocabulary is localised, in a tradition as far as I know limited to upper Egypt, initially carried out by family members but now more often contracted to artisans (who increasingly have art school training). Typically, a *hajj* painting will state the year of the pilgrimage, and show the means of transport – usually an aircraft but in earlier examples a boat – with scenes of the sacred sites at Mecca. To undertake the pilgrimage is a religious obligation for those able to do so, but expensive and requiring at least ten to fourteen days' absence from home. Although most pilgrims are men the *hajj* is not restricted to them. Within a village such paintings carry considerable significance, and the wealth of a village may incidentally be estimated (by outsiders like myself) from the number of such paintings indicating the number of dwellers who have made the journey. The significance, though, is not monetary but religious, and both rests on and reproduces a common value structure and cultural framework which links Muslim communities from Indonesia to West Africa, and through migration within Western cities. I do not want to sentimentalise this – to me inevitably an exotic other – but find interesting that the transnational model of globalisation is not the only one available. Returning to Dosh, there is a café selling Egyptian-produced cola as well as tea, outside which I, my research assistant and interpreter Louise, the guide, the driver and a courteous but perhaps bemused police corporal sat for a while, joined by two adolescent boys. I cannot begin to say what detail of lives took place behind the small wooden doors of the houses, any more than I know what goes on inside the old apartment blocks of Barcelona. I take away only visual impressions, aware that they are incomplete, and what is vouched to me in translated conversations.

further demolition. The term used for these clearances is *esponjamiento*, a loosening of the weave of the urban fabric. As Degen reports, 'After the Olympics the council made it a priority to upgrade these neighbourhoods to improve the quality of life of their residents but also to open them up for tourism' (Degen, 2004: 139). The square outside MACBA is a favourite for skateboarders in the evening and dog walkers in the early morning. But while plans to build a luxury hotel on the new Ramblas remain on hold, the area is now less ethnically mixed. Rent controls and a requirement to rehouse people dispossessed by improvements have retained elements of immigrant and older populations, but young professionals have moved into the refurbished apartment blocks with their new amenities and high service charges. As a young planner commented, 'people like us are moving in' (in Degen, 2002: 28).

Not all the small art galleries and boutiques around MACBA have remained, and the area is still a zone of authentic tourism experience. Yet one change to the built environment in particular denotes a shift in the city's image: new apartment blocks have no balconies, or balconies so shallow as to be unable to support the traditional load of birds in cages, plants, tins of oil, gas canisters and people in their vests. That image states a Latin city – a poor city – while the new façades create the image of a cool, northern city – an affluent city.

A Universal Forum of Cultures, 2004

The abolition of the transitional space of the balcony as an outdoor (public) extension of indoor (domestic) space is evidence of the city's realignment in a global economy. Degen draws attention to tensions between residents and night-time drinkers in La Ribera, another newly recoded neighbourhood (Degen, 2004: 140). She concludes:

> While Barcelona's regeneration initially grew out of a social-democratic project and local consensus, its success in entering the global tourist circuit means that economic concerns are beginning to overpower local needs. This is one of the consequences of becoming part of a 'global cities network', and the increased competition between locations where local policies are designed to satisfy a global market.
>
> (Degen, 2004: 141)

The networks and flows in which the city participates include two arcs: from Valencia through Barcelona to Montpellier, Marseilles and Genova; and from Barcelona to Madrid and Lisbon. A further sign of the city's realignment to these arcs is the designation of the old working-class neighbourhood of Poble Nou as a knowledge-quarter for small businesses in graphic design, media and communications. A strategy of clearance is overt in the most recent phase of the city's redevelopment for the Universal Forum of Cultures in 2004. This covers the waterfront from the Olympic village area and twin towers – described by Diane Ghirardo as out of scale and representing 'the typical exploitation of urban centres and beachfront property throughout Western cities' (Ghirardo, 1996: 200) – to the Besós river. It is in that far reach, now called Diagonal Mar, previously a zone of electricity and water works, railway lines and redundant industrial sites, that Avenida Diagonal, which bisects *Eixample*, meets the sea. The adjacent area was a site of informal housing, residual nineteenth-century workers' terraces and artists' studios in redundant factories.

Officials described the project as a design problem, and a brochure elucidates sweeping intentions:

> A new residential district is planned . . . These dwellings will be designed and priced to attract young people and thus provide a shot in the arm for an area where the population has aged . . . Rehabilitation work is also planned on the blighted bordering areas . . . The 'La Catalana' area is located in the Sant Adrià

del Besós municipality . . . a residential neighbourhood which has gradually lost population over the years. There are currently just 80 families living on the site. They will have to be relocated . . . Part of the land is publicly owned, thus making execution of the scheme much simpler.

(Ajuntament de Barcelona, 1999: 20)

The Forum was a cultural successor to the Olympics, and Beatriz Garcia sees Barcelona as an event-led city (Garcia, 2002), but it was equally a means to rebrand the city as the hub of the western Mediterranean arc. A guide to Barcelona's architecture states that 'Private enterprise has also been unusually active . . . in the Diagonal Mar area, under the impetus of the holding of the World Forum of Cultures' (González and Lacuesta, 2002: 106) while a key site was assigned to North American developers for a gated park surrounded by high-rise, high-rent apartments. There is a large mall nearby, and eleven new hotels to service the new convention centre next to the Forum building.

While the city's marketing in the 1990s emphasised cultural tourism linked to Picasso, Miró and other well-known artists and architects, by the end of the decade it had added an approach based in business conventions and the search for continuing inward investment, using a more market-led ethos in context of global competition. Construction around the Forum site reflects this under the guise, in 2004, of a cultural Olympics drawing on the universal humanism of postwar cultural internationalism. The Forum emphasises cultural diversity, citing the Universal Declaration of Human Rights and appealing to a global public in an aim to promote 'discussion on topics affecting all of human kind' while raising questions that 'will shape the world's development' and involving the public in 'this process of examination and reflection' (Universal Forum of Cultures, 2003: 16). The brochure continues:

> Pavilions will be opened where visitors can sample and enjoy cuisine and crafts from hundreds of different cultures, and visit exhibits of innovative international fashion and design. The World Festival of Arts will give special attention to artistic creations that emphasize the mix and artistic collaboration between different cultures, particularly those which have often been at odds with one another.
>
> (Universal Forum of Cultures, nd: 32)

The claim to resolve conflict can be put beside Zukin's argument (1995) that culture conceals conflict and redirects demands into consumption. It can also be put, with the high-rise apartment blocks in their landscaped park, beside Cerdà's *Theory of Urban Construction*, published in 1859:

> Our whole social edifice . . . reposes on the base of property. Its free use . . . is indefinite and unlimited, as long as it does not redound to the harm of a third

party; but if it reaches this situation, property is an act of usurpation, and freedom, a licence. Hence the restrictions which all municipal ordinances place on the use of property.

(Cerdà, 1859: §520, in Soria y Puig, 1999: 417)

Cerdà argued that planning can integrate commercial expansion with public benefit through regulation. The Forum eclipses this to set other precedents in the pursuit of a globalised affluence resting more on deregulation. I am left wondering whose city is re-presented under the guise of culture.

End-thoughts

The marketing of cities for tourism and as sites for the financial services industries relies on the idea of a city as a concentration of resources and attractions. This is a traditional view (see Chapter 1) and may have declining currency in a period of globalisation and corporate decentring as companies out-source key provisions such as accounting and publicity so that competitors are serviced by the same bank or public relations agency. Added to this is a creative dimension of advertising which re-presents an oil company as beyond petroleum, for instance. The idea, then, of a unified company with a coherent structure and purpose housed in a prestigious headquarters building may be defunct, and perhaps with it in time the idea of a city as a unified entity reflecting the unified citizen-subject of modernity and encapsulated in a unified branding image. This may render the place-identity strategies of city marketing redundant. Perhaps a symbolic economy of entirely immaterial production no longer relying on material places will discard the idea of the city altogether.

But perhaps there will also be a recognition that ordinary cities, as Jennifer Robinson calls them (2006), are as cultured as spectacular cities, interrupting a history of identification of power with the places in which it is exercised. I end this chapter and section by quoting Robinson:

> A reconceptualisation of urban modernity might begin by interrogating Western forms of modernity. For these are as much a product of dynamics from beyond the West as its own ingenuity, including primitivist-inspired borrowings, actual resource extraction and political innovations learned in colonial governance . . . But western modernity's projection of itself as the generative source of creativity relies on forgetting these circulations and borrowings.
>
> (Robinson, 2006: 19)

Within this terrain are also alternative modernisms, such as that represented by Hassan Fathy (see Chapter 1); and no doubt alternative postmodernisms.

Notes for further reading

Chapters 3 and 4

Much of Walter Benjamin's writing is available in translation, most editions reprinted several times – I give the most recent in the bibliography. For his *Arcades* project I recommend the Harvard edition (1999) which is a new translation; his writing on Paris and Baudelaire (see Chapter 3) is in *Charles Baudelaire* (1997); his essay 'The Author as Producer' of 1934 (relevant to Chapters 3, 4 and 5) is in *Understanding Brecht* (1998). Among many critical commentaries I suggest Esther Leslie's *Walter Benjamin: Overpowering Conformism* (2000) as the most incisive, and Graeme Gilloch's *Myth and Metropolis: Walter Benjamin and the City* (1996) as emphasising urban contexts and theories. The standard critical work specific to the *Arcades* project is Susan Buck-Morss's *The Dialectics of Seeing* (1991), an impressive work. Alex Coles offers a study of Benjamin's aesthetics in 'The Ruin and the House of Porosity', in an edition of *de-, dis-, ex-* (vol. 3, 1999). Pessoa's writing is available in English in a Penguin edition (2001). There are several editions of Baudelaire's poems in English, in most of which the translator's style reinflects the content. Although outside my scope here I recommend Herbert Marcuse's essay on French literature in the 1940s (Marcuse, 1998: 199–214) for its brief comment on Baudelaire and the culture of intimacy. Several longish extracts of Baudelaire's criticism are contained in *Art in Theory 1815–1900* (Harrison, Wood and Gaiger, 1998). Its companion volume, *Art in Theory, 1900–1992: An Anthology of Changing Ideas* (Harrison and Wood, 1992), covers the movements of modernism and conceptualism. Given the male bias in most art history I recommend *Feminism-Art-Theory: An Anthology 1968– 2000* (Robinson, 2001) as antidote. Other sources on feminism and cultural production include Jo Anna Isaak's *Feminism & Contemporary Art* (1996) and Katy Deepwell's *New Feminist Art Criticism* (1995). These could be read with bel hooks's *Art on My Mind* (1995), from which I cite an extract in case study 8.2.

Sharon Zukin's *The Culture of Cities* (1995) is relevant to this section and Chapter 5, as is her chapter 'Cultural Strategies of Economic Development and the Hegemony of Vision' in *The Urbanization of Injustice*, edited by Andy Merrifield and Erik Swyngedouw (1996) – in my view her clearest exposition of her arguments on symbolic economies and cultural re-coding. This could be read beside the last section of Doreen Massey's *Space, Place and Gender* (1994), and the essays in *Imagining Cities*, edited by Sallie Westwood and John Williams (1997). Marsha Meskimmon's *Engendering the City: Women Artists and Urban Spaces* (1997) gives insights on women's places in the urban spaces of modernist art. Hal Foster's *The Return of the Real* (1996) includes an essay on the contemporary artist as ethnographer which extends the terrain I have set out both to emerging art practices in the 1990s and to a juxtaposition of methods in art and social science fieldwork (see also Chapter 5). Bourdieu's ideas are best encountered in his own writing, as in *Distinction* (1984) and *The Field of Cultural Production* (1993). His conversation with artist Hans Haacke, *Free Exchange* (Bourdieu and Haacke, 1995), offers a lucid discussion on art's critical function and could be read with Jacques Rancière's *The Politics of Aesthetics* (2004).

Nike Culture by Robert Goldman and Stephen Papson (1998) critiques the manufacture of consumerist culture. The literature of cultural consumption is extensive and includes studies of its institutional spaces such as *Art Apart*, edited by Marcia Pointon (1994) with essays on Tate (see Chapter 2) and the Museum of Modern Art, New York. The essays in *The Politics of Consumption*, edited by Martin Daunton and Matthew Hilton (2001), reconsider the consumption of luxuries and necessities in the past two centuries, with a chapter asking 'What Is Rum?' in relation to the politics of the French Revolution. At another polarity in this field is *Tourism Mobilities*, edited by Mimi Sheller and John Urry (2004), with an essay by Monica Degen on Barcelona which I cite in Chapter 4; and Bella Dicks's *Culture on Display* (2003) on the visible production of visitability.

City Visions, edited by David Bell and Azzedine Haddour (2000), observes and critiques socio-cultural conditions from mobility to cosmopolitanism. *City of Quarters*, edited by David Bell and Mark Jayne (2004), includes a section on 'Production and Consumption'. Jon Binnie et al. provide studies of Montréal, East London, Amsterdam, Birmingham and Manchester's Gay Village in *Cosmopolitan Urbanism* (2006), which also includes a potentially seminal essay on mongrel cities by Leonie Sandercock. I cite an extract from one of its chapters on Singapore in case study 6.2. I recommend Iain Borden's *Skateboarding, Space and the City* (2001) as a narrative of skateboarding as everyday spatial production (or embodied architecture as Borden terms it). David Crouch's work on allotments (2003 – case study 3.2) covers another area of everyday cultural production. These titles touch on the everyday cultures of cities, and the works by Henri Lefebvre cited in the notes for further reading to Chapters 1 and 2 remain relevant here. Linking the everyday with mid-twentieth-century cultural criticism I suggest, finally, Siegfried Kracauer's *The Mass Ornament* (1995) with essays on the hotel lobby and the Linden Arcade (Kaisergalerie) in Berlin – an equivalent of those of Paris, built in 1873, modernised in 1928 and destroyed by bombing in 1944.

Part Three

Culture industries and cultural policies

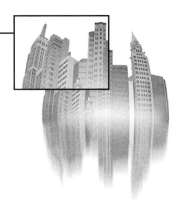

5 The culture industries

Learning objectives

- To understand what the culture industries are and how they operate
- To differentiate between terms and definitions used in different geographical and ideological contexts
- To test the chapter's arguments through the case of new urbanism

Introduction

In Chapter 3 I looked at cultural movements in context of large cities, and in Chapter 4 at the rebranding of a city through its cultural infrastructure. The demise of manufacturing in the affluent world means that the cultural sector now has a key role in the economies of many cities. In this chapter I examine the culture industries, borrowing the term from Sharon Zukin (1995) and finding that definitions vary (Evans, 2001; Scott, 2000; Landry and Bianchini, 1995; Myerscough, 1988) as well as ideological perspectives. I ask whether the rise of the culture industries represents a culturation of the economy (Warde, 2002) or an economisation of culture, and link today's use of the terms *culture industries*, or *cultural and creative industries*, to that of the culture industry in critical theory in the 1940s. I turn next to Disney, critiqued by Michael Sorkin (1992) and Sharon Zukin (1995), looking at its venture into real estate at Celebration, Florida (MacCannell, 1999; Wood, 2002). Finally, I ask whether state support for culture mirrors Disney's instrumentalism. In case studies I cite a news report of an art project at Abu Graib in Iraq, and a text on Polish factory film clubs by Marysia Lewandowska and Neil Cummings; I illustrate a billboard work by Sheffield-based artists Mel Jordan and Andy Hewitt exhibited at the Guangzhou triennale in 2005.

The culture industries

Context

Art and literature are regarded in the modern period in the West as fields of individual creativity, yet creative individuals reach a public via mediation. A writer who wins a major prize does so through ability but there would be little to celebrate without the processes which produce the prize, its promotion, a market for books and the profession of writer. Andy Warhol's idea that everyone can be famous for a while seems to refute this but I know of it only because Warhol was an artworld star. The relations of individuals in the cultural sector to its institutions are vital to success, and artists, writers, designers and craftspeople increasingly need to know the map of the sector in which they function, the policy agendas which influence it and opportunities provided by interactions within it – whether defined as the arts and heritage (Myerscough, 1988) or to include film, broadcasting, advertising, design and mass media (Landry and Bianchini, 1995). Yet the arts constitute an unusual model for an industry in that most artists and performers are self-employed, and it is not surprising that the sector tends more often to be broadly defined to include the immaterial production of public relations and media sectors, and at times the material production of luxury goods in the fashion trades.

The term *culture industries* denotes, then, an extensive and diverse grouping of elite and mass cultural production and distribution. A distinctive feature is that parts of it – the arts and public broadcasting in particular – are state enterprises managed for public benefit. There is a continuing debate as to how far the arts should be funded in this way or regarded as businesses (Jowell, 2005), and as to whether a public-benefit ethos can or should survive in a period of market deregulation. In the UK the Arts Council was established in the 1940s to widen access to the visual and performing arts within the new welfare state (Pearson, 1982: 53–55). The first Arts Minister, Jenny Lee, was appointed in the Wilson administration in 1964 in the Department of Works (Willett, 1967: 202–203). Today the Department for Culture, Media and Sport (DCMS) oversees the Arts Council and devises policies for broadcasting and heritage.

The fine arts are represented in DCMS mapping of the cultural economy only by the art and antiques market, which may seem remote from the image of an artist's studio as site of autonomous creativity. But this is not a new situation. Rembrandt's paintings were produced for the art market in the Dutch Republic in the seventeenth century, and the auction houses which sell his work now are heirs to the market in which he was engaged for most of his working life. What is new is that his failure rather than success in that market, his bankruptcy in 1656 and withdrawal to domestic life, are subsumed in a mythicised idea of the artist as outsider (see Chapter 3) to increase the market value of his work, overlooking his real empathy with the poor in Dutch society (Molyneux, 2001: 69–87). For governments, though, the role of the arts in urban economies has replaced faith in aesthetic quality as a civilising influence. Bianchini et al. state 'the best strategic programme for improving the quality of life might well turn out to be based on developing a coherent and wide-ranging arts and cultural policy' (Bianchini et al., 1988: 10), a sentiment repeated in an Arts Council booklet, *An Urban Renaissance*: 'The arts are crowd-pullers. People find themselves drawn to places which are vibrant and alive' (Arts Council, 1989: np). Andrew Kelly writes that 'For cities to develop . . . on the national and international scale, they need to become culturally vibrant' (Kelly, 2001: 16). Arts districts attract professional-class bohemians (Wilson, 2000: 240–249), and one of them, Hoxton in London, is 'the capital's trendiest area' (Evans, 2004: 71). Writing on Berlin after German reunification Graeme Evans notes that the city relies for its cultural reputation on visual artists while rises in property values have forced them to find new sites of production elsewhere (Evans, 2001: 175–176, citing Kotowski and Frohling, 1993). In Berlin this has led unusually to artists' studio and housing provision by the city. In most cities the appeal of past cultural production and present cultural distribution is regarded as adequate to feed a symbolic economy.

Nicholas Pearson writes that 'Artists make art, but the culture is the product of an interaction between artists, their work, patrons, buyers, commissioners, educators, historians and critics' (Pearson, 1982: 97). Zukin notes that some artists are politically engaged, others 'co-opted into property redevelopment projects as beneficiaries' (Zukin, 1995: 23) aiding a city's claim to cultural hegemony. This may be an outdated view as the culture industry moves from traditional art forms towards new media and a fusion of art, design and consumption, as in the designer bars which line the canal side in Manchester's Castlefield district. The ironic taste of cultural recoding is illustrated by the names of two adjacent bars under the railway arches in Castlefield: Revolution and Fat Cats. The irony is not confined to Britain: in Yerevan, Armenia, a steel hammer and sickle announces a restaurant (see Chapter 7).

A heightened perception of a city fuels the economies of tourism and leisure, depending on a critical mass of cultural attractions. While relatively few people are directly employed in theatres or art galleries, they draw in audiences many of whom stay in hotels, eat in restaurants, drink in bars, use taxis and go shopping in high streets and malls, to create employment – though much of it is insecure and badly paid (Loftman and Nevin, 1998) and in a globalised economy much of the profit goes elsewhere than the city in which it is made. Bearing in mind the discussion of culture, creativity and economic sectors in the previous chapter, it might be appropriate to view city marketing itself as a cultural sector, not only the cultural attractions it sells. It is clear, however, that the word *culture* in the culture (or cultural and creative) industries denotes culture as the arts, or extensions of that to immaterial production and the manufacture of image, rather than to cultures as the diverse ways of life of multiple urban publics. At times highly mediated versions of this, as in folkloric shows, may be marketed as part of a city's image, but the daily lives of dwellers are no more marketable than accessible to the aestheticising gaze.

Definitions and categories

To look in more detail at definitions: Allen Scott (2000) includes fashion, furniture, jewellery, musical instruments, toys and sporting goods, handbags, hats, perfume manufacture, broadcast media, motion film and architectural services in a list of cultural-product industries in the USA (Scott, 2000: 9). This contrasts with John Myerscough's (1988) definition, kept to areas in which the development and dissemination of original artistic ideas is primary, in the visual and performing arts and museum sectors (Rodgers, 1989: 1). Charles Landry and Franco Bianchini (Landry and Bianchini, 1995; Landry, 2000) include popular music, arts festivals, new media, design and cultural tourism. Their definition of creativity is 'a way of getting rid of rigid preconceptions and of opening ourselves up to complex phenomena which cannot always be dealt with in a strictly logical manner' (Landry and Bianchini, 1995: 10). Landry and Bianchini are advocates for the cultural industries (as they call them), seeing cultural production and consumption as solutions to deindustrialisation and social unease:

> Today many of the world's cities face tough periods of transition. Old industries are disappearing, as value-added is created less through what we manufacture and more through the application of new knowledge to products, processes and services.
>
> (Landry and Bianchini, 1995: 3)

As this shows, their definition of the creative sector extends to the knowledge economy and its role in knowledge economy quarters for information and communication technology and design sectors, parallel to the designation of cultural quarters. Calling for a holistic approach in urban policy they cite examples of projects from festivals to new public spaces, street entertainment to mixed-use zoning, temporary exhibitions to happy hours in inner-city bars, and community mosaics to city rebranding. While Landry and Bianchini see popular music and film as able to bring together members of diverse ethnic groups – 'New ideas can be generated through cultural crossovers' (Landry and Bianchini, 1995: 25) – Evans regards their work (citing Landry, 2000) as reflecting a 'professionalisation and bureaucratization of both cultural and other public policy realms and decision-making structures' (Evans, 2001: 277). I tend to agree with Evans and wonder at the ease with which models of cultural renewal can be sold through cultural industries consultancy regardless of localised conditions.

To return to definitions of the cultural sector, Evans divides it into three, which I summarise:

- cultural industries (print and broadcast media, music, design, art markets, and digital media)
- cultural tourism (arts venues, heritage sites, events and festivals)
- arts amenities (subsidised and local arts, and civic provision for the public good).

Each requires a different kind of consideration, as he says,

> the first a more traditional concern for the means of production . . . ; the second, environmental planning that seeks to balance carrying capacity and visitor flows, transport and scale of facilities against cultural policy goals . . . ; and finally arts-as-amenity which places civic arts resources, facilities and activities in a local/subregional planning context.
>
> (Evans, 2001: 140)

Evans differentiates the production of cultural objects from service areas such as heritage culture while emphasising the importance of local conditions: while 'traditional local and neighbourhood arts and cultural amenities . . . present few planning problems . . . new supply-led developments . . . increasingly "distort" the basis of traditional public leisure planning' (Evans, 2001: 142), citing multiplex cinemas and late-night clubs. Scott has a less value-laden approach, seeking to demonstrate the size and diversity of cultural industries, investigate the production of high and low culture in the conditions of global capitalism and analyse the spatial logics of the cultural industries in different sectors.

Evans and Scott take an evidence-based approach, increasingly the position taken by the UK government in evaluating its cultural expenditure. But there is a difficulty in that quantitative data such as the numbers of gallery visits or theatre attendances give no indication of the quality of experience. If it is claimed, for instance (fashionably in the 1990s), that more people go to museums and galleries than to football matches this reflects only that galleries and museums are mainly free while football matches are expensive, were then prone to crowd violence and can be viewed on television. If art is experienced personally, it is probably not viable to measure the emotive impact of a gallery visit, leaving only the multiplying effect of spin-off employment or role of cultural institutions in redevelopment to be estimated.

Case study 5.1

Sgt Lynne Steely, 'Abu Graib Detainees Enter Art Contest'

Abu Graib, Iraq (Army News Service, October 6th, 2005):

Concrete bunkers, strategically placed within the confines of Abu Graib prison for detainee protection, turned into works of art when juvenile detainees were offered the challenge to paint them in the form of a contest.

The detainees were given paint and supplies provided by the 306th Military Police Battalion's Repair and Utilities section to decorate the bunkers, and a few theme ideas, such as 'a united Iraq' to get them started.

Capt. Jim Allen, compound commander, said he came up with the idea as a way to keep the juvenile minds occupied and to give them something to focus on.

'The juveniles become bored very easily,' said Allen. 'We're always trying to think of new activities for them.'

'we also wanted to tie the contest in with the beginning of Ramadan to help get them mentally set for the holiday,' added Allen. Ramadan . . . is a holy month where Muslims offer peace and treat other people with kindness and respect.

The bunkers were painted in a myriad of colours, in several designs and patterns. Some displayed phrases in English or Arabic. One bunker read, 'Help we to new Iraq. We need freedom,' amongst several painted flowers.

The contest concluded Oct. 3 when the judges determined a first-, second- and third-place winner. Judges for the contest included Lt. Col, John Hussey, 306th MP Battalion commander; Col. Bernard Flynn, Forward Operating Base Abu Graib commander; William Ivey, Task Force-134 deputy commander; and Lt. Col. Michael Blahovec, 18th Military Police Brigade deputy commander.

First-place winners received a large edition of the Qur'an (the holy book of Islam) and seven nights of comedy movies. Second-place also received a large Qur'an and one night of comedy movies. Each of the third-place winners received seven days of special meals that included sweet snacks and treats.

The detainees enjoyed the contest and are looking forward to another one, according to the 18th MP Brigade Arabic translator.

Another contest is planned for after the Ramadan holiday, said Allen.

(available at http://www4.army.mil/ocpa/read.php?story_id_key=8024)

Note

I am grateful to Nicola Kirkham for forwarding this item to me from UK Indymedia. Lynne Steely is a Sergeant in the 18th Military Police Brigade Public Affairs Office at Camp Victory, Iraq.

Comment

I include this item because it illustrates the extent to which the expediency of culture (Yúdice, 2003) is normalised. Just as art projects are used as solutions or holding operations for domestic social problems and badges of respectability for property developers, they migrate into the foreign adventures of the super-power. Of course, it is better that detainees have something to do, and painting walls is better than being subjected to degrading treatment of more obvious kinds. I put it that way because being told the narrative of one's national identity by a soldier in an occupying army is degrading, too. But paint is better than bullets. Yet I hardly know what to say, as someone who in a previous phase of my career used to argue for the arts in non-gallery settings. I'll try description: there is a landscape on the side of a hut. Houses and trees are dotted over hills. There is a road, and a ridge with trees in the foreground. Five-pointed stars are painted around the door-frame. I remember the debates on community-based arts in Britain in the 1980s, and the claim that aesthetic criteria were irrelevant. This claim marginalised community arts and led to shortfalls in funding, and to a divide between community artists and those, usually closer to the mainstream, who taught in art education and used this as a network of opportunity. I remember several projects which seemed to serve the interests of neither the peripatetic community artist nor the colonised community. There is a long history, too, of art in prisons and health services where it has a therapeutic and expressive quality. In that situation part of the therapy is in making personal choices, like whether to do it or not. At Abu Graib the incentive is global movies or sweets. But what strikes me most is how the case demonstrates a normalisation of art's policing function, its utility in mopping up the time of people the system regards as transgressive. It is touching that the commander credits Muslims with offering peace and treating other people with kindness and respect (during Ramadan). The image which accompanies the text shows a mesh fence laced with razor wire.

A cultural economy

Public policy in the industrialised world has undergone something of a cultural turn since the 1980s as the public sector has seen the arts as levering private-sector investment in urban areas, and meeting social needs. Scott's underlying purpose is to test claims 'currently fashionable in some social science circles' that economic reality is embedded in cultural reality (Scott, 2000: ix). He shows how local environments shape the cultural industries in different ways, and compares the media-based cultural sector of Los Angeles to the more traditional cultural economy of Paris:

> if the cultural products of Los Angeles are today relatively dominant on world markets whereas those of Paris are relatively restricted, it is nevertheless possible to envisage a . . . rectification of this situation in future. More generally, there are important reasons for supposing that cultural production in capitalism in the twenty-first century may be rather more polycentric, in geographic terms, than it was in the twentieth.
>
> (Scott, 2000: 170)

Scott shows the rise of the film industry in Los Angeles (Scott, 2000: 175–176, tables 11.1 to 11.3), and finds a cultural sector employing a total of 412,392 people in Los Angeles County in 1996. He concludes, 'culture-producing sectors are now moving to the very forefront of capitalist development and growth' (Scott, 2000: 204), while high and low culture are fused in the production of image as commodity. Fredric Jameson, whom Scott cites, proposes similarly that 'If modernism thought of itself as a prodigious revolution in cultural production, . . . postmodernism thinks of itself as a renewal of production as such after a long period of ossification and dwelling among dead monuments' (Jameson, 1991: 313).

Scott's work can be compared to Zukin's on loft living in SoHo (1989) and employment patterns of artists and actors in New York (1995: 142–152). But Zukin observes that it is difficult to measure arts employment: 'While it is hard to gauge employment in the visual arts because artists hold so many different jobs, it is tricky to estimate employment in the theatre because jobs are not continuous' (Zukin, 1995: 145). She exposes ties between real estate speculation and cultural institutions (Zukin, 1995: 109–142), and is critical of the encroachment of the private sector on public space (Zukin, 1995: 133–142), if less grounded in the actualities of art-working than artist Martha Rosler whose project *If You Lived Here* (1989–91) used the site of SoHo's art-space to foreground issues of eviction and gentrification in the district after its recoding. But if it is easy to see urban redevelopment as a Disneyfication of urban space (Zukin, 1991; 1995: 49–78) and the culture industry as dominated by global conglomerates, Scott sees small-scale, artisanal modes of production in traditional and new media – 'dense networks of specialized but complementary producers clustered together in industrial districts whose roots extend deeply into the fabric of some of the world's major cities' (Scott, 2000: 205) – as drivers of urban regeneration.

It emerges that different definitions of the cultural industries mirror different investigative tasks. Landry and Bianchini promote creativity as new ways of thinking about local government and planning. Zukin's agenda is a reassertion of the place of social justice in urban discourse (which I read as a key value in her writing). Scott and Evans ask how the cultural sector operates, using detailed analyses of specific sites. But as Mark Jayne comments, many studies offer 'generalizeable accounts which have been grounded in case studies of such areas within cities which are near the top of the urban hierarchy' (Jayne, 2000: 13, citing Ley and Olds, 1988; Boyle and Hughes, 1991; Soja, 1996). Many of them are un-theorised as well, as much research in the cultural policy and planning sector is contracted by clients seeking reassuring answers to questions of regeneration.

Cultural goods

In buying a pair of branded trainers, what is consumed is the brand name more than the product (Klein, 2000). Sociologists have read into this the formation of personal and group identities. There is historical evidence of a link between consumption and self-perception, and not all of it is negative – the department store, for example, offered women freedom from domesticity (Rendell, 2002: 17, citing *inter alia* Wilson, 1992; Dowling, 1993; Ryan, 1994). Cinemas, clubs and bars offer publics today opportunities to, so to speak, be themselves with others; but I wonder if consumers play the market at its own game, or are its dupes. I accept that shopping is central to postmodern life (Shields, 1992) but wonder if shoppers are as active in shaping their lives as is claimed (Ritzer, 1999), operating within the finite limits of what the market brings them while the quasi-magical status of the commodity is affirmed. Minjoo Oh and Jorge Arditi summarise a complex field: 'Writing about shopping shifted the focus from the production to the consumption aspects of economic practice' (Oh and Arditi, 2000: 71, citing Edgell, Hetherington and Warde, 1996; Lury, 1996; Lee, 1993; Featherstone, 1991). Conrad Lodziak refuses the prevailing position that shopping is postmodern empowerment, seeing a manipulation of needs 'that occurs as a consequence of the alienation of labour and our dependence on employment', a material rather than ideological form of manipulation which 'takes place prior to the secondary . . . ideological manipulations of advertising and the media' (Lodziak, 2002: 94). He reconsiders the split of basic from non-survival needs (in which consumerism reflects surplus wants) to argue that consumerism is a response to new survival needs:

> our lives remain structured by the need to survive . . . the manipulation of this need by the social system *imposes* . . . a pattern of consumption relevant to survival. . . . Furthermore, so-called unnecessary consumption . . . can also be best understood, not so much as compulsory, but as manipulated by the social system.
>
> (Lodziak, 2002: 96)

I read Lodziak's argument in terms of the production of psychological needs in a society of competitive productivity. As he says, 'For many, the recuperative needs that are generated by employment ensure that whatever free time there is is given over to rest and relaxation' (Lodziak, 2002: 101). He cites Theodor Adorno and Max Horkheimer: 'the individual who is thoroughly weary must use his weariness as energy for his surrender to the collective power which wears him out' (Adorno and Horkheimer, 1997: 152–153, in Lodziak, 2002: 101); and comments that the entertainment which fills this manufactured need is a means to procure audiences for advertising, re-enforcing the compulsion to consume in a vicious circle. So, beyond the consumption pattern of the leisure class described by Thorsten Veblen (1970) – wealth is demonstrated by spending time in non-necessary ways – cultural experience is democratised through mass availability but its consumption is a response to the pressures of life which produce new needs met, with built-in lack of satisfaction as in all commodity fields, by cultural goods.

Culture and economy are imbricated when the presentation of consumption is cultivated. The question is whether economic practices are generally cultural. Alan Warde is critical of this idea:

> some contemporary accounts conclude that because economic practice is meaningful, it is thereby cultural. In turn, this leads to the suggestion that no conceptual distinction can be sustained between the economic and the cultural. However, to hold such a view entails that there can be no answer to the question 'Is the economy becoming more cultural?', for unless we can distinguish clearly between economic activity and cultural activity, culturalization would be unidentifiable.
>
> (Warde, 2002: 185)

Warde's idea could be compared to John Allen's argument that all economic sectors are creative (2002, see Chapter 4), and he identifies the origin of the problem in a 'profligate use of the term "cultural" in the recent social scientific literature' (Warde, 2002: 185) to cover almost anything.

Warde differentiates two trends: an increase in cultural inflection of economies; and an increase in economic inflection of culture. Cultural studies has brought ordinary life into the frame of its investigations but a by-product may have been a crossover of the term culture in its anthropological sense to economy. Warde observes a double value judgement at work:

> Most recent analyses of the relationship between culture and the economy assume that the boundaries between them are collapsing . . . Yet there are two alternative causal claims. The former, 'culturalization', implies that culture has become more central in economic relations, a process usually implicitly considered good because no one is against more culture. The latter suggests

that economic relations are now more central to culture and, because this might be the result of commodification or rationalization, this might be bad.

(Warde, 2002: 185–186)

Warde takes Williams's three categories of culture – intellectual development, a way of life, the arts – in *Keywords* (1976 – see Chapter 2) and applies them in reconsidering culture and economy, asking if culture is becoming integrated in economic activities, citing the information society. He asks if culture as a way of life increasingly impacts economic relations; and if the arts are a new area of productivity with direct and indirect links to materiality. Citing Celia Lury (1996) he takes her work on consumption to affirm all three directions, and recapitulates: 'It is not merely the products of the culture industries supplied for final consumption, but their role in affecting advertising, marketing and packaging which are significant in introducing culture to the economy' (Warde, 2002: 188). Warde sees the impact of these weak relations of culture and economy as 'negligible' for a majority of workers when, for instance, a company's structure is creatively re-engineered (Warde, 2002: 189). He adds, 'while Lury has isolated those trends and processes relevant to a culturalization of consumption, the effects are, as yet, comparatively limited, with only a small part of all production governed by such imperatives' (Warde, 2002: 193).

Warde's conclusion is that cultural experiences have a larger impact today than in the past but this implies 'neither the cultural saturation of everyday life nor the attenuation of the logic of economic practice' (Warde, 2002: 198). He argues that cultural expenditure is a minor element of household expenditure, and ends: 'the economy has a remarkable capacity to recuperate and discipline the outputs of cultural taste' (Warde, 2002: 199). This is an antidote to advocacy for the cultural and creative industries as drivers of urban regeneration (Landry, 2000) and touches incidentally on another aspect of the cultural turn in urban policy, its police function. If, as Michael Dorland (2000: 145) argues, the modern state administered the needs and wants of citizens for their happiness, cultural policy is an instrument of managing conformity. This leads me to critical theory.

The culture industry

For Marx economy was the base, and culture the superstructure, of a society's organisation. Changes in culture followed changes in economic conditions, but a separation of base and superstructure is questioned now. When barriers not only between art forms but also between the arts and fashion, mass media and consumption are dissolved it is unhelpful to separate the cultural from the economic. But the problem is more structural in that – this is the key point –

Case study 5.2

From Marysia Lewandowska and Neil Cummings, *Enthusiasts*

Founded in 1964, the amateur film club AKF 'Kalejdoskop' in Turek is a good example of such a tension [between autonomy and the production of propaganda]. It is enough to say that the Propaganda Department . . . and the Department of Culture . . . were among its 'patrons'. In exchange for patronage by the local authorities, and financial backing . . . the film makers were encouraged to produce films about local events, and sites of historical interest. War related themes were among those most appreciated. During the Cold War period, the citizens of the former Eastern Block lived in permanent fear of violent confrontation. According to communist propaganda the West was ready to strike at any time, any day. The 'Kalejdoskop' club released a film called Veto . . . [which] ends with a 'souvenir' photo of American soldiers posing on top of a tank adorned with Merry Christmas.

Who exactly was the amateur film maker? Jerzy Petz explains . . . 'As the very name indicates it is a person, who makes films out of inner need – to please himself, to relax after work, to make his life richer . . . Some documented their private lives, and their films were just keepsakes . . . There was also another, a very distinctive approach: that of an amateur who wants to share his world with others, who want to represent his life, his passions, his doubts, and the circumstances of his emotional life.' Some amateur artists in communist Poland also enjoyed a relative freedom – their passion allowed them to explore subjects which were impossible to include in officially approved film and television production: homosexuality, youth subcultures, promiscuity, laziness, satire, etc. Amateurs also voiced critical opinions about their surrounding reality . . . In their films, made in the 60s and 70s, Poland does not always appear to be a socialist Arcadia, because depressing cityscapes, littered with misconceived architectural developments are far from idyllic . . . Amateur film makers were some of the most sensitive to these problems and responded, by registering the worst aspects of living in socialist cities.

(Lewandowska and Cummings, 2004: 94–97)

Comment

The research which led to the exhibition and publication cited here shows a self-organised film industry running parallel to the cultural apparatus of the socialist bloc.

I say *industry* in the sense that the amateur film-makers worked in a variety of genres. This puts the autonomy claimed for modern culture in perspective since the autonomy here is organisational rather than aesthetic. Neil Cummings, asked why he took an interest in this, replied that amateur film-making, or the work of the enthusiast and hobbyist, 'opens onto a range of interests and experiences generally invisible amongst the relentless flow of the state sponsored, or professionally mediated' (Lewandowska and Cummings, 2004: 18). In context of a view that the culture industries purvey the enforcement of capitalism, it is important to observe instances of cultural formations outside that frame. Kate Soper writes that, while there is critical work on consumerism in the period of the War on Terror, there is little on 'what a counter- or post-consumerist order might look like, what alternative seductions to McDonaldization it might evoke' (Soper, 2005: 7). Perhaps a self-organising society is a facet of that post-consumerist world, or a world in which cultural policy is determined by cultural producers. This would implement – as it does in this case – a practical ownership of the means of production by the workers. It would be a case of the author as producer, as Benjamin puts it in an address to a group of communist writers in Paris in 1934 (Benjamin, 1998: 85–104), arguing that it is not enough to depict the desired transformation of society, but necessary to realise it in the means of cultural production. Benjamin cites the case of the Soviet press, in which readers are also contributors:

> we see that the vast melting-down process of which I spoke not only destroys the conventional separation between genres, between writer and poet, scholar and popularizer, but that it questions even the separation between author and reader. The press is the most decisive point of reference for this process, and that is why any consideration of the author as producer must extend to and include the press.
>
> (Benjamin, 2003: 90)

The press, like photography – which he also cites in this text – and film, and experimental theatre, is a technical medium offering the possibility to dissolve the aura associated with art through the rites of observance which take place in high culture's institutions. An implication of this paper is that the status of a professional artist or writer dissolves when members of the mass society are cultural producers. Perhaps, still, the amateur film-makers of Polish factories are workers who have taken over the cultural means of production to tell their own stories and, implicitly at least, organise their own society.

the supposed base and supposed superstructure are in a relation of mutual conditioning. Culture re-presents and reconstructs the economy while the economy reconstructs and represents culture dialectically, just as the subject is conditioned by her or his environment while acting on it.

These statements extend from Marx's eleventh 'Thesis on Feuerbach' (1845, in Marx, 1947: 197–199) in which he sketches a philosophy of practice. For Ernst Fischer, 'Philosophy without practice dissolves very easily into air . . . practice without philosophy turns into myopic, mindless practicism' (Fischer, 1973: 157). Such comments refuse an opposition of culture and economy as well. Reconsidering this may lead me to reconsider the meaning of what I have called non-necessary consumption, and in turn may imply that the idea of consumption as identity formation can be understood as resting on a new type of necessity (as Lodziak argues, above) provided by the culture industries. At this point, though, I need to explain why I use this term in place of the more commonly encountered *cultural* (and *creative*) *industries*. So far in this chapter I have moved between such terms and a looser cultural sector. But to use the term 'culture industry' is to refer to critical theory (Adorno, 1991). Zukin's *culture industries* seems to fuse this position with acceptance of the diversity of sub-sectors within cultural production today compared to the primacy of the movies and broadcasting as vehicles of control in the interests of capital in Adorno's writing.

The term *culture industry* derives, then, from the work of the Frankfurt Institute for Social Research (or Frankfurt School). Adorno explains why he and Horkheimer adopted the term in the 1940s, in North America after fleeing the Nazi regime in Germany, in his essay 'Culture Industry Reconsidered':

> In our drafts we spoke of 'mass culture'. We replaced that expression with 'culture industry' in order to exclude from the outset the interpretation agreeable to its advocates: that it is a matter of something like a culture that arises spontaneously from the masses themselves, the contemporary form of popular art. From the latter the culture industry must be distinguished in the extreme.
> (Adorno, 1991: 85)

Their antagonism is at odds with Walter Benjamin's enthusiasm for film as a vehicle of public imagination in his essay 'The Work of Art in an Age of Technical Reproducibility' (*Das Kunstwerk im Zeitalter seiner technischen Reproduzierbarkeit*, 1935 – see Benjamin, 1970: 219–255; Leslie, 2000: 130–167). For Benjamin the audience identifies with the labour of the actor, knowing her or his work to be repetitive. It foresees divergent plot lines or endings and an implication is that an audience able to do this can imagine alternative narratives of social formation. For Benjamin film offers hope; for Adorno the movies deceive. It is not a contradiction: Benjamin looked at experimental film in Germany and Russia; Adorno wrote about mass culture, drawing attention to the extent to which culture standardises its representation of life: 'Films, radio and magazines make up a system which is uniform as a whole and in every part' (Adorno and Horkheimer, 1997: 120). It is the world of Hollywood after the advent of sound:

> Real life is becoming indistinguishable from the movies. The sound film, far surpassing the theatre of illusion, leaves no room for imagination or reflection on the part of the audience.
>
> (Adorno and Horkheimer, 1997: 126)

It deceives just as astrology reproduces the imperatives of capital such as the separation of work from pleasure (Adorno, 1994: 71–77). Adorno objects that the culture industry splits production from reception to construct passive audiences unable to affect its products, while 'How this mentality might be changed is excluded throughout' (Adorno, 1991: 86). In 'The schema of mass culture' he writes:

> All mass culture is fundamentally adaptation . . . The pre-digested quality of the product prevails, justifies itself and establishes itself all the more firmly in so far as it constantly refers to those who cannot digest anything not already pre-digested. It is baby-food: permanent self-reflection based upon the infantile compulsion towards the repetition of needs which it creates in the first place. Traditional cultural goods are treated in just the same way.
>
> (Adorno, 1991: 58)

Mass culture masks conflict to promote a conflict-free society, designed to reproduce the economic system which invented it. As he says, 'The dream industry does not so much fabricate the dreams of the customers as introduce the dreams of the suppliers among the people' (Adorno, 1991: 80). And in *Dialectic of Enlightenment*:

> The culture industry perpetually cheats its consumers of what it perpetually promises. The promissory note which . . . it draws on pleasure is endlessly prolonged; the promise, which is actually all the spectacle consists of, is illusory: all it actually confirms is that the real point will never be reached, that the diner must be satisfied with the menu.
>
> (Adorno and Horkheimer, 1997: 139)

I now turn to Disney as a case in which Adorno's observations can be brought into a contemporary frame.

Disney Celebration

Michael Sorkin writes that Disneyland was designed as a utopia. He cites its early publicity:

> Disneyland will be based upon and dedicated to the ideals, the dreams, and the hard facts that have created America [*sic*]. And it will be uniquely equipped to dramatize these dreams and facts and send them forth as a source of courage and inspiration to all the world.
>
> (cited in Sorkin, 1992: 206)

I am unsure what facts are intended here but the prescription has a resonance with the political and military rhetoric of the Cold War and its successor the War on Terror (Buck-Morss, 2002a). Sorkin illustrates his paper with a photograph of the sky above Disney World as the only picture he is allowed to reproduce in 'the first copyrighted urban environment in history' (Sorkin, 1992: 207).

Among the ancestors of Disneyland (founded in 1955) and Disney World are the World Fairs, or International Expositions, of the nineteenth century and their more recent successors, and the malls of the late twentieth century (Crawford, 1992). The Great Exhibition at Crystal Palace, London, in 1851 represented British manufactures, as the national pavilions at recent World Fairs – such as Expo '92, Seville (Harvey, 1996: 59–69) – showcase technological innovation interrelated to national identity. Penelope Harvey remarks that national exhibits have been used increasingly to 'demonstrate cultural homogeneity' when national cultures are 'quite openly presented as heterogeneous and fluid communities' (Harvey, 1996: 59). Sorkin observes that the popularity of exhibitions in which the land-scapes of tropical environments were represented in winter gardens declined when rail travel made it possible to travel to exotic places (Sorkin, 1992: 210). International expositions exhibited the ethnography of colonial possessions (Coombes, 1994a; 1994b) and this, too, may be eclipsed by the advent of travel by ocean liner. This raises a question as to why anyone would want, in a period of cheap air travel, to visit Disneyland. Sorkin's answer is that 'dislocation is central' to it (Sorkin, 1992: 210); and that Disney's themed environments are 'intensifications of the present' (Sorkin, 1992: 216). He remarks that the symbolic rendition of a global marketplace in World Fairs is reduced here to 'the pith of a haiku', that 'the goods at Disneyland represent the degree zero of commodity signification', while Disney World's 'national' pavilions 'groan with knick-knacks' which are 'not simply emblems of participation in the enterprise of the higher, global, shopping, they are stand-ins for the act of travel itself, ersatz souvenirs' (Sorkin, 1992: 216.). Easy. Lite.

Zukin reads Disneyland and Disney World as 'the alter ego and the collective fantasy of American society' (Zukin, 1995: 49). This is contextualised by Disney's extensive media interests which distribute and normalise its brand of enter-tainment. She sees the visual coherence offered by its environments as creating 'a public culture of civility and security that recalls a world long left behind' (Zukin, 1995: 52). Edward Soja views new urbanism in the work of Andreas Duany and Elizabeth Plater-Zyberk at Seaside, Florida, and Playa Vista, Los Angeles, as, 'a peculiar postmodern combination of historical urban nostalgia and present-day post-suburbia' (Soja, 2000: 248). But there is a darker side: Zukin notes that in Disney's sites there are no guns, homeless people or drug abuse, but also, 'Without installing a visibly repressive political authority, Disney World imposes order on unruly, heterogeneous populations – tourist hordes and the work force that caters to them – and makes them grateful' (Zukin, 1995: 52).

The Disney dream became real estate in Celebration, Florida, a development of houses, shops, a school, a theatre and gardens with white picket fences surrounded by wetlands and forest. Andrew Wood notes that Celebration is an extension of Walt Disney's 'plan to turn the community into a hybrid exposition/ theme park/profit centre whose exhibits just happened to be residents' (Wood, 2002: 190); and Emilio Tadini reads Celebration as quintessentially postmodern in its eclecticism, remarking that 'What we have here is the illustration of a myth . . . a certain America is dreaming of itself – and seems so satisfied . . . it decides to realise it, life-size, and to present it to the world as the True American Dream' (Tadini, 1997: 59). Celebration has a town hall but a Town Manager employed by Disney instead of an elected mayor. Its residents sign up to a 'Declaration of Covenants' prescribing in detail the visual appearance and social use of the built environment, to the level of the colour of curtains, shrubs that can be planted and frequency of garage sales. The houses are in colonial style, in pastel colours, the designs drawn from Charleston, South Carolina, and Savannah, Georgia, both heritage sites. They are also connected to a cable network providing residents with what are seen as the tools of community building. As Wood observes, 'Celebration homes represent a haven of instantaneous communication and high-bandwidth data transfer . . . Residents seeking to visit an age of organic ritual and cyclical return do so upon a foundation of buried technology.' (Wood, 2002: 200). The visible impression, however, is dominated by the white picket fences (also on the town logo). Journalists Catherine Collins and Douglas Frantz, who lived in Celebration for a year, write that 'Nowhere was that backward vision more evident than in the *Celebration Pattern Book*, which was assembled to serve as the design bible for the town' (Frantz and Collins, 1999: 64), which they contrast to Asher Benjamin's *The American Builder's Companion*, a pattern book from the prewar period which they see, rightly or not, denoting a design consensus not an imposition.

Dean MacCannell notices a peculiarity in the design of houses at Celebration: a panopticon-like visual path from porch to back yard, as denoting a nostalgia for central authority. He continues, 'The panopticon dwelling eliminates the possibility of discovering anything that might disconfirm the hypothesis of deep spiritual harmony' (MacCannell, 1999: 113). He examines Celebration in relation to the surveillance mentality of post-industrial urban dystopianism, and likens it to the visible face of recent television police dramas. The advent of reality television gives this alignment of community, safety and old-fashioned values a further twist. But Wood draws out the intricate and contradictory screening and fusion of public and private spaces in Celebration, noting from the Pattern Book

Case study 5.3

Figure 5.1 *Hewitt + Jordan,* **The Neo-Imperial Function,** *billboard, Guangzhou triennial exhibition, 2005 (photo courtesy of the artists)*

Comment

Hewitt + Jordan produced a series of billboard works between 2004 and 2006 on the functions of public art. The project began when they were hired by Public Art Forum (PAF – a UK network of public art agents and officers) to do something creative at its 2003 annual conference. They offered themselves as the prize in a raffle, the winner being able to commission them to make a work within the specified budget provided by PAF. The outcome was the first billboard, in Sheffield in 2004, stating 'The economic function of public art is to increase the value of private property' (Hewitt + Jordan, 2004: 53). The artists write:

> Our practice is defined by our political and social engagement; it is process-based and its outcomes a result of the method and research carried out. We work with sites of debates and usually in conjunction with others . . . We intend to contest accepted hierarchical structures . . . as well as attempting to reveal the ideological agendas that exist within art itself.
>
> (Hewitt + Jordan, 2004: 7)

They contest art's functionality in a market-led system, and its function as a badge of respectability in urban redevelopment schemes. A second billboard, in Leeds in 2005 as part of the project *Vitrine* curated by Pippa Hale and Kerry Harker, says 'The social function of public art is to subject us to civic behaviour'; the third – a banner for the Venice Biennale – states '*La funzione estetica dell'arte pubblica é quella di rendere naturali distinzioni sociali*' (The aesthetic function of public art is to codify social distinctions as natural). A fourth, on a billboard in Hackney commissioned by the curating partnership B+B (Sophie Hope and Sarah Carrington), stated 'The function of public art for regeneration is to sex up the control of the under-classes'. The texts are in each case devised by Hewitt and Jordan with collaborating artist Dave Beech. This one foregrounds the neo-imperialist function and began as a billboard text in Birmingham's Chinese Quarter, was photographed there and then taken as a photograph to China, where it was publicly displayed and rephotographed for exhibition at the second Guangzhou Triennale (in a section curated by Gavin Wade). It has since been rerephotographed for billboard exhibition again in Birmingham. In Mandarin, it says: 'The neo-imperialist function of public art is to clear a path for aggressive economic expansion' (Hewitt + Jordan, 2005: np).

that back-yard spaces are reserved (beyond the panoptic gaze but outdoors) for family pursuits (which do not presumably include the conception of new family members) while front porches – found in most houses – are community zones, where residents see and are seen by others passing by. There is frequent reference to community-building in Celebration's published material (see Wood, 2002: 193–194), yet it is hard to resist the interpretation that this is, in effect, a means of control. Wood writes:

> The domestic realm is produced by the façade through the astute invocation of scale and detail. Written into these elements is the orderliness and linear assurances of proportion and control. Within the public sphere that is enacted through the screen of architecture, one finds a world that is visible and thus subject to institutional regulation.
>
> (Wood, 2002: 193)

The ethos of control has not always worked – residents objected to the one-class school reminiscent of a nineteenth-century small town but inadequate for today's educational needs – but journalist Betina Drew sums up Celebration's visible charm: 'on a lovely August day people were outside enjoying their porches and eager to talk. The nearness of the houses to the street made me feel easy and included, and I was surprised at how intensely familiar it all felt' (Drew, 1998: 175). Yet she writes that although Celebration is the first town to be more private than public, while the boundaries of these realms are invisible, there is an exclusivity:

> The Disney-owned parks and sidewalks look like parks and sidewalks anywhere; and officers from the Osceola County sheriff's office ride the streets in patrol cars. But the pool is for Celebration residents and their guests only, and the officer I spoke to . . . was there off-duty at Celebration's request . . . a bonded, highly trained employee to supplement Disney's security, which patrols less conspicuously at night. Finally, there is the town hall . . . almost universally disliked, the unease it provokes undoubtedly more than architectural: There is no town government as Americans [*sic*] know it.
>
> (Drew, 1998: 179)

Gerard Delanty argues that the theme of loss of community runs deep in Western culture (Delanty, 2003: 15–21), and this is taken up by Wood when he reads Celebration as an exercise in an inherently flawed nostalgia. The flaw is in the experiment's reproduction of dominant patterns of spatial gendering: 'Disney reacts to a perceived need in American [*sic*] society through the rhetorical and architectural construction of nostalgic discourse that subtly affirms problematic assumptions of gender' (Wood, 2002: 199). Delanty notes that liberalism alone among modern ideologies is not constructed around the social but on individualism, and that the control-ethos of Celebration is dependent on fears which arise from a breakdown of the social realm.

If 'The interpretation of the New Urban community and the panoptic house at Celebration . . . constitutes a suppression, even an erasure, of human difference except as a (very) few demographic categories recognized by corporate community planners' (MacCannell, 1999: 114), then Celebration seems to be an updated form of the white, North American suburb in which outsiders and people of colour are seen as suspect, described by Richard Sennett in *The Uses of Disorder* (1970). It may reproduce a privileging of expertise, translated from the realm of rational planning in the 1950s into the power of the Town Manager and his [*sic*] global employer. Wood writes that 'The Victorian reflects an age that defined the "expert" and celebrated the potential for technology to execute more efficiently the process of social definition' (Wood, 2002: 201). That definition did not include revolt, and used cultural and social reform to ensure future stability regardless of deeper and structural difficulties which now reappear in what is called social exclusion and the emergence of an underclass – terms applied by the excluders rather than the excluded (who do not constitute a unified class at all).

Conclusion: cultural instrumentalism

If the culture industries – Zukin's term – serve urban redevelopment and the production of symbolic economies (see Chapter 4), and include both small- and medium-scale enterprises and transnational conglomerates, a question arises as to whether the arts are tools for social engineering – that is, thinking of the cultural turn in urban policy and Warde's questioning of the relation between culture and economics, whether the arts have undergone a utilitarian turn as instruments of other areas of policy. To an extent the arts were seen in the days of the welfare state as educationally useful in the nineteenth-century manner of improving the minds of the lower classes (see Chapter 2). But since the transition from arts administration to arts management in the 1980s, arts professionals have been invited by governments to seek funding through compliance to public policy agendas. The result of the success of their lobbying is that the arts were seen in the 1980s and 1990s as able to resolve social and economic problems. Yet the case for cultural utility is not evidence-based; and the problems the arts are required to address are produced by other areas of government policy and the market operations to which policies are aligned. I am disturbed by this for three reasons: first, if culture masks socio-economic problems it normalises them and displaces attention from appropriate interventions; second, the cultural takes the place of the political – a cultural economy in place of political economy – because it is non-contentious, or, as Zukin writes, the arts 'create a new landscape that diverts attention from old political battles' (Zukin, 1996b: 227–228); and third, a mask of universal value derived from aesthetics constructs new mechanisms of stabil-isation and control in what is, in widening social and economic divisions, an inherently unstable situation. It is a bit like facing climate change by introducing new colour schemes for heavily polluting power stations.

To give an indication of the extent of the utilitarian turn in the arts I cite Gary O. Larson's report for the National Endowment for the Arts (NEA), quoted by George Yúdice in *The Expediency of Culture*:

> No longer restricted to the sanctioned arenas of culture, the arts would be literally suffused throughout the civic structure, finding a home in a variety of community service and economic development activities – from youth programs and crime prevention to job training and race relations – far afield from the traditional aesthetic functions of the arts. This extended role for culture can also be seen in many new partners that arts organizations have taken on in recent years.
>
> (Larson, 1997: 127–128, cited in Yúdice, 2003: 11)

Yúdice notes that arts managers must 'produce and distribute the producers of art and culture, who in turn deliver communities or consumers' (Yúdice, 2003: 13), which I think puts impossible demands on artists. The myth of the modern artist may credit him or her with special insight (Kuspit, 1993) but socio-economic problems need more than such mastery (as it tends to be), and I do not subscribe to the myth anyway. The difficulties faced need resources, policies and empowerment, not art. They also need imagination but this is one of the qualities the disaffected already have, along with an ability for self-organisation (Roschelle and Wright, 2003). In this context I recall Henri Lefebvre's theory of moments of liberation, the idea that in the routines of daily life people have moments of a sudden and piercing perception of reality which are liberating (Shields, 1999: 58–64). I recall Zukin's comment on the fashion industry which recycles motifs from street-life into designer goods: 'The cacophony of demands for justice is translated into a coherent demand for jeans' (Zukin, 1995: 9). Perhaps the most dangerous facet of the culture industry is, in conclusion, not that it introduces the dreams of the suppliers to the market as Adorno says but that it re-presents society as a unified, coherent entity – like an artwork.

6 **Cities of culture**

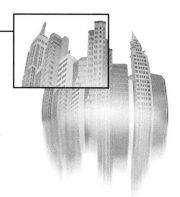

Learning objectives

- To investigate the European Union programme of Cities and Capitals of Culture
- To inquire what values the programme represents
- To compare cases within the programme

Introduction

In Chapter 5 I investigated the definition and critical evaluation of the cultural sector. In this chapter I turn to the specific case of European Cities of Culture to see how the sector is mediated by transnational political agendas and the interests of specific cities. I set out the European Commission's policy for cultural cities, noting past and future cases and some of the issues raised. I then look at Glasgow (City of Culture in 1990) and Bergen (one of nine cities selected for 2000), noting issues of ownership and sustainability. Finally, I compare the programme's aims with those of a UNESCO report, 'Our Creative Diversity' (1996), which sees cultural provision as a means to empowerment. Given the focus on institutional structures in the main text of this chapter, case studies are used to expand its scope. I quote from a narrative of cultural work within the struggle for national liberation in Eritrea in the 1980s by Amrit Wilson; and an account of the cultural milieu of Singapore by Serene Tan and Brenda S. A. Yeoh in which the authors observe the role of fictional narratives in culturally led aspirations to cosmopolitanism. Then, coming back to a European City of Culture but to a recent artist-initiated project, I illustrate a spoof street sign in Weimar by Portuguese-Brazilian artist Daniela Brasil, exhorting citizens to kiss.

Cultural cities

The European Union (EU) and Cities of Culture

The programme for European Cities of Culture, or Capitals of Culture from 2000 (ECOC), was devised by the European Commission (EC) to 'contribute to bringing the peoples of Europe together'. Beatriz Garcia describes it as pragmatic in 'adapting to the needs and demands of those cities hosting it rather than imposing a prefigured model of urban cultural policy' (Garcia, 2005: 841). The list of nominated cities (table 6.1) shows a geographical balance reflecting EU membership. In 1999 the method of nomination for Capitals of Culture from 2005 onwards was refined to involve a panel of seven independent experts from the national cultural sector in the selection process. National governments nonetheless retain the key voice in the nomination:

> Each member state shall submit, in turn, its nomination of one or more cities to the European Parliament, the Council, the Commission and the Committee of the Regions. This nomination shall be submitted no later than four years before the event in question is due to begin and may be accompanied by a recommendation from the Member State concerned.
>
> (Official Journal of the European Union,
> decision 649/2005/EC, 13 April 2005, Article 1, ¶2)

Member states are currently required to begin selection six years in advance of the year in which their nominated city will be a Capital of Culture, with a meeting of the panel five years ahead to begin shortlisting from submissions.

Liverpool will be a Capital of Culture in 2008, after a competition in which Belfast, Newcastle-Gateshead, Bristol, Cardiff, Oxford and Birmingham were among the contenders. In most cases submissions brought together a coalition representing elements of existing cultural life, and some unsuccessful bidders will develop parts of the proposed programme by other means. Bristol's image as a legible city (Kelly, 2001), for instance, has already produced new signage and the installation of internet kiosks across the city centre.

The enlargement of the European Union (EU) to include members of the former East bloc is reflected in dual nominations of states for the years 2007 to 2019 (tables 6.2 and 6.3). In formal terms, the selection criteria include:

> a European dimension, which should foster multilateral cooperation between cultural operators at all levels, highlight the richness of cultural diversity and bring the common aspects of European cultures to the fore

Table 6.1 *European Cities of Culture, 1985–99, and Capitals of Culture, 2000–06*

1985	Athens	1993	Antwerp
1986	Florence	1994	Lisbon
1987	Amsterdam	1995	Luxemburg
1988	Berlin (West)	1996	Copenhagen
1989	Paris	1997	Thessaloniki
1990	Glasgow	1998	Stockholm
1991	Dublin	1999	Weimar
1992	Madrid		

2000	Avignon, Bergen, Bologna, Brussels, Helsinki, Cracow, Reykjavik, Prague, Santiago de Compostella
2001	Porto, Rotterdam
2002	Bruges, Salamanca
2003	Graz
2004	Genova, Lille
2005	Cork
2006	Patras

Table 6.2 *European Capitals of Culture, 2007–10*

2007	Luxemburg, Sibiu
2008	Liverpool, Stavanger
2009	Vilnius, Linz
2010	Pecs, Essen or Görlitz

Table 6.3 *European Capitals of Culture host states, 2011–19*

2011	Finland, Estonia
2012	Portugal, Slovenia
2013	France, Slovakia
2014	Sweden, Latvia
2015	Belgium, Czech Republic
2016	Spain, Poland
2017	Denmark, Cyprus
2018	Netherlands, Malta
2019	Italy

and

> a citizenship dimension, which should raise the interest of the citizens living
> in the city and from other countries and be an integral part of the long term
> cultural development of the city.
>
> (Decision of the European Parliament and of the Council,
> 30.5.2005; COM(2005) 209 final, 2.2D)

Such criteria are open to interpretation but imply the presence of a critical mass
of cultural institutions with appropriately diverse programmes, a notable cultural
history and cityscape, a commitment to European identity, provision for wide
public benefit and viable projections of growth. Liverpool has in these respects
Tate Liverpool, the Walker Art Gallery and Bluecoat Gallery; a history of popular
music and Rope Walks cultural quarter (Gilmore, 2004); the buildings of the
Victorian waterfront and a museum of the slave trade on which the city's wealth
was built; and a projection of economic recovery from a market-led city management team. Being Capital of Culture in 2008 will increase Liverpool's tourism
income and lend the city the prestige required to attract inward investment.

The second Capital of Culture in 2008 will be Stavanger, Norway. Its project
has strong support from the Commune (city authority), follows Bergen's year
as a Capital of Culture in 2000 (see below) and was announced in 2004 with
international conferences on 'The City and the Grassroots' and 'The City and the
Work of Art', and the installation of forty sculptures by Antony Gormley – each
a cast-iron cast of the artist's body – throughout the city, including in domestic
and retail settings as well as public spaces. While Liverpool's recent history is
of deindustrialisation Stavanger's is of growth based on the oil and natural gas
industries. The city seeks to rival Bergen as a cultural hub and centre for fjord
tourism, seeing the cultural programme as contributing to its symbolic economy.
If the intentions expressed in 2004 are carried through, its year of cultural
celebration will also involve a grass-roots level of input to the programme's
definition.

Mapping culture

The record of the European Parliament's approval for a new system of
nominations in 2005 states that the programme 'is geared towards highlighting
the wealth, diversity and shared characteristics of European cultures and towards
contributing to improving European citizens' mutual knowledge' (Official Journal
of the European Union, decision 649/2005/EC, 13 April 2005, ¶2). It requires
cities to ensure that their event will 'be sustainable and be an integral part of the
long term cultural development of the city' (Decision of the European Parliament

and of the Council, 2005/102 (COD), Article 3, ¶3b). It enables cities to participate in a Europe-wide network to produce another mapping of European urbanisation than that of national capitals or commercial centres, and to compete for status globally. Judith Kapferer writes of 'a shift in perspective which has the contemporary city by-passing the national State to run up against the global politics and economics of a capital expansion of increasingly unfettered transnational conglomerates' (Kapferer, 2003: 31). Of the forty-one cities which were or will be Cities or Capitals of Culture from 1985 to 2010 only fourteen are national capitals.

The idea of alternative mapping raises issues of whether the programme is aligned with efforts to achieve world-city status – not incompatible with regional cultural provision as shown by Barcelona (Chapter 4); and whether the benefits of cultural festivities are evenly distributed. As with symbolic economies the question is whose city is culturally reconfigured for whom. A shortage of evidence as to the impacts of such projects (Evans, 2005; Bailey, Miles and Stark, 2004; Lim, 1993) fuels scepticism while the evidence that exists portrays a failure to deliver aims, or define them in ways the delivery of which can be demonstrated. Graeme Evans notes that 'distributive effects and regeneration objectives . . . are generally under-achieved – or they are not sustained' (Evans, 2005: 975). Tim Hall remarks: 'Moments of civic transformation tend to get portrayed in overly simplistic terms as seamless and unproblematic. The reality is much more messy' (Hall, 2004: 63, cited in Evans, 2005: 961). This is heightened when cities devise themes which are broad enough to allow all potential stakeholders to be involved and specific enough for marketing agencies to use them in campaigns, drawing on selective histories which tend towards the seamless.

Cultural themes

The nine cities nominated for the year 2000 adopted themes from the geographical (Bergen: coasts and waterways) to the technological (Avignon: history and technology) and artistic (Reykjavik: voices of the earth) (table 6.4), while all competed global millennium celebrations.

From observations of projects in Avignon, Bergen and Bologna, Kapferer concludes that, while they gave prominence to the arts and facilitated inter-city connections, 'crucially, the European Culture City celebrated . . . the pre-eminence of trade as defining the urban experience' (Kapferer, 2003: 40). She alludes to histories of mercantile families, the Hanseatic League and Elizabethan piracy as contexts for economic competition, and as suggesting networks which crossed national boundaries in an earlier Europe. These references suggest

Table 6.4 *Thematic descriptions of Capitals of Culture for 2000*

Avignon	Tecnomade (history and technology)
Bergen	Coasts and waterways
Bologna	Communications
Brussels	The city past and present
Cracow	Codex Calixtinus (musical performances)
Helsinki	Café 9 (performances and installations)
Prague	Telelink (technology)
Reykjavik	Voices of Europe (choral music)
Santiago de Compostela	Faces of the Earth (cartography)

(adapted from Kapferer, 2003: 38)

romanticisation, but I want to draw out an inference parallel to that of the global city (Sassen, 1991). For Saskia Sassen the global city is not a material city in a conventional sense but unified by communications links between financial industries districts in cities such as London, New York and Tokyo. These districts have little social, cultural or economic contact with geographically adjacent urban neighbourhoods. Sassen regards the global city as a site of post-industrial production which participates in all economic sectors and builds markets for the circulation of its products through deregulation, internationalisation, innovation and creative risk (Sassen, 1991: 124). This produces a map which has little connection to the boundaries of nation-states but is coherent as a map of post-industrial production. Sassen concludes that a small number of cities, national capitals like London and Tokyo, or commercial centres like New York, have concentrations of key functions in the new economy (Sassen, 1991: 164–165), and are able to keep them while other cities are underrepresented in this loop regardless of profiles which rise for other reasons (Sassen, 1991: 190). The EC faces this competition in seeking to establish forward-looking economies in its key cities, most of which are not world cities and do not have concentrations of financial industry services but were made national or regional capitals in previous phases of national or regional formation. Culture is thus a key means to remap the status of such cities or build the importance of financial centres such as Paris and Frankfurt, Milan and Barcelona, hitherto outside the global city circuit. It is also a means to raise the profile of relatively small cities, such as Weimar, which have major cultural histories intertwined with political significance (Eckardt, 2006). Trade histories subsumed under the banner of heritage and re-presented in cultural tourism become adjuncts to such efforts, but may draw attention away from intractable, underlying problems. Evans sees the Cities of Culture programme as diverting policy and resources for structural economic change into culturally led redevelopment; this bypasses national governments while 'the use of culture as a conduit for the branding of the "European Project" has added fuel to culture city competition, whilst . . . celebrating an official version of the European urban renaissance' (Evans, 2003: 426). Garcia, citing Evans, notes that this is contested (Myerscough, 1994) but adds, 'there is little question of the [EC] programme's effect on increasing city competitiveness and promoting culture-led regeneration agendas in an expanding Europe' (Garcia, 2005: 843). In Avignon cultural events are managed by a private company linked to services including water and cable television (le Galés, 1993: 197). But Kapferer ends with a nice picture:

> the people of Avignon, Bergen and Bologna still imagine the city as theirs. The city itself continues to be manifested as the image and the reality of capital accumulation and distribution, of course. But the walls of the Cities of Culture . . . potently symbolise the local terrain of a still distinctive time and place . . . the inhabitants continue to lead daily lives of joy and sadness, comedy and tragedy, hope and despair, getting and spending on a human scale.
>
> (Kapferer, 2003: 42)

Cities of Culture

The UK experience

Franco Bianchini argues that 'Glasgow, Bradford and Birmingham, cities severely hit by the decline of manufacturing industry during the recessions of the 1970s and early 1980s, achieved substantial changes in image through their use of cultural policy' (Bianchini, 1993: 17). Of these, Glasgow has the highest cultural profile; Birmingham is England's second city and Bradford remains economically depressed. Postcards on sale in Birmingham are captioned 'Birmingham City of Culture' and feature the city's Victorian civic architecture and recent public art – such as Centenary Square, a 3-acre space with decorative paving by Tess Jaray, public art and new street furniture, and an International Convention Centre (Hall, 1997; Miles, 1998; Hall and Hubbard, 2001; Hall and Smith, 2005). The city centre is a central business district on the Burgess concentric ring model (Chapter 1), an impression enforced by its isolation within a road network, but re-presented as a cultural district. There are new office buildings in the adjacent cultural quarter around the Ikon Gallery, their workforce supplying the customers for nearby designer bars and cafés along the restored canals.

Glasgow has more mystique than Birmingham. Garcia reports 55.3 per cent positive ratings for Glasgow's image (16.8 per cent negative) but, supporting the idea that image counts more than material conditions, ratings of 27.8 per cent positive (48.4 per cent negative) for its quality of life (Garcia, 2005: 853). Garcia also cites several press headlines from the year of preparation for City of Culture which contributed to Glasgow's external perception and were used in city marketing: 'The ugly duckling of Europe has turned into a swan', for example (*Los Angeles Herald Examiner*, 27 August 1989, cited in Garcia, 2005: 855). She cites press coverage of, or claims for, the city's ability after 1990 to attract cultural tourism; and concludes that economic concerns have risen on agendas while those of participation have fallen (Garcia, 2005: 856). Press reports are not data but feed into a symbolic economy and influence perceptions which impact future growth prospects. The lesson of Birmingham's culturally led redevelopment is similarly mixed. Patrick Loftman and Brendan Nevin (1998; 2003) view the economic benefits of Centenary Square as mainly temporary, low-paid employment in service sectors such as catering. Many of the food outlets are global franchises rather than contributors to local or regional economies. Despite the efforts to create a visual distinctiveness in Centenary Square (now encroached on by commercial spectacles), the only trace of a genuinely local cultural life I could see on a visit in the summer of 2005 was a stall selling artefacts painted by a retired canal barge painter under the sign LOCAL ARTIST.

The issue of local and non-local economies is raised, too, by the mobility of capital in setting up new centres of production. Bob Jessop argues that cities compete in a zero-sum game in which mobile factors of production move from one site to another but do not increase (Jessop, 1998: 96). The difficulty is that money flows quickly out of a local or regional economy into a transnational economy. Jessop points out that entrepreneurialism in a neo-liberal climate is far from socially neutral, transferring resources and power 'from localities to firms, from local collective consumption to investment, and from immobile to mobile forms' (Jessop, 1998: 97). Hall argues that Birmingham's cultural remodelling 'opened up a series of centres and peripheries according to their compatibility with the identities of these wider spaces [of cultural valorisation]', but that alignment with redevelopment 'had the effect of distancing it from any potentially radical or subversive meanings' (Hall, 1997: 218). Loftman and Nevin see Birmingham's culturally led policies as a diversion of scarce public-sector resources 'away from essential services such as education and housing – particularly affecting the city's most disadvantaged communities' (Loftman and Nevin, 1998: 147). These experiences from sites of economic need addressed as cultural aspiration may be instructive for the Capitals of Culture programme.

Glasgow 1990

Peter Booth and Robin Boyle quote Michael Donelly, previously Assistant Curator at the People's Palace: '1990 was a year when an intellectually bankrupt and brutally undemocratic administration projected its mediocre image onto the city and ordered us to adore it' (in Booth and Boyle, 1993: 21). Aram Eisenschitz says that during Glasgow's year as City of Culture, 'its indigenous socialist culture was systematically downgraded' (Eisenschitz, 1997: 171, citing McLay, 1990). Whether or not this was the case, the story of prestigious projects sidelining local networks is familiar. Julia Gonzalez writes of the eclipse of local cultural scenes by the Guggenheim in Bilbao, and that a definition of quality of life used in the city's cultural policy lacks a social dimension: 'a major problem for a considerable number of local citizens' (Gonzalez, 1993: 83). She notes contested allegiances in Bilbao's cultural policy between 'expressing Basque cultural identity, as a vehicle for stimulating the citizens' political participation, and as an element of the quality of life needed for economic revitalisation' (Gonzalez, 1993: 88).

My recollection of Glasgow in 1990 is that the city had a reputation for experimental theatre, and several vanguard artists' groups and artist-run spaces as well as the reflected reputation of internationally exhibited graduates from its art school; that there had been two decades of community arts development with gable-end murals, and a garden festival in 1988 on a redundant industrial site by the Clyde; and that the opening of new cultural institutions such as the Burrell Collection – a private collection of fine, applied and world arts – added to the sense that the city had more to offer a cultural tourist than Edinburgh. The city's nineteenth-century housing stock was renovated rather than demolished to make way for roads as in the 1970s, while pockets of deprivation remained near the shining cultural landmarks and conference hotels. Booth and Boyle remark that within living memory Glasgow had been 'an industrial city of world stand-ing' (Booth and Boyle, 1993: 24) but the signs of this were relegated to heritage status.

Booth and Boyle also say that the development of cultural and redevelopment policies in Glasgow is confusing, dominated by deindustrialisation, social problems and a legacy of poor-quality social housing in peripheral blocks (Booth and Boyle, 1993: 27). An exception was the Glasgow Eastern Areas Renewal Project, following the 1978 Inner Urban Areas Act. But, as Michael Pacione comments, 'a wide-ranging national and local attack on urban problems failed to materialize' (Pacione, 1997: 27). In Glasgow industry as well as housing was encouraged to move to outer-city sites so that, as Booth and Boyle put it, 'The outcome of Glasgow's urban renewal was not simply spatially divisive; it was also sterile' (Booth and Boyle, 1993: 28). In the 1990s housing policy shifted to repopulation of the city centre, producing gentrification and a cultural renewal based on emergent cultural attractions such as the Tron Theatre and nearby Café Gandolfi. Booth and Boyle see the success of the inner-city Merchant Quarter as unforeseen but enabled through public-sector resourcing from the Scottish

Case study 6.1

From Amrit Wilson, 'Eritrea: The Experience of Creating a New Culture'

In the bases of the liberated zone, everyday life is like a series of wonders. An area may be so well-camouflaged that the casual observer would think it unpopulated . . . Total self reliance is a crucial aspect of Eritrean culture. Nothing is wasted. In converted containers, surgeons perform the most highly skilled operations . . . [They] are not foreign experts but people who a few years ago may have been rural housewives or nomads . . .

It was in a place like this in 1987, Orota, that I witnessed and took part in the celebration of International Womens Day and encountered Eritrean revolutionary art . . . Rooms were cleared, halls prepared and constructed, stages erected, suwa, the local millet wine, prepared in large quantities . . .

Around midday, on happening to pass a local bakery, a building camouflaged like so many others into near invisibility, we are invited in. To our surprise, a play is going on inside. When our eyes get accustomed to the darkness we see men, women and children, many in traditional dress, sitting in rows watching the show, commenting and drinking suwa . . .

In the workplace, with a space cleared for a stage and floursacks as curtains, the workers perform themselves and their community. The spirit of such an event with the commitment and resourcefulness of the participants is hard to imagine in Britain today.

One play is about the role and sufferings of a mother. A woman is harassed by two Ethiopian spies who demand information about her sons. They beat her and throw her to the ground, but she refuses to break down. The security forces move in and she is imprisoned. Her sons, freedom fighters, return and kill the spies but the woman dies in prison. Women holding their babies brush tears from their eyes. It is a story close to everyone's lives.

(Wilson, 1988: 202–203)

Comment

Amrit Wilson begins this account from personal experience among the Eritrean Peoples Liberation Front (EPLF) in the 1980s with a summary of Eritrean history, from Italian colonisation in the 1890s to federation with Ethiopia at US insistence in 1942, armed struggle from 1961 onwards and a continuation of repression after regime change in

Ethiopia. She notes that 35 per cent of EPLF fighters are women, that reforms include setting up village assemblies and ending of forced marriages, and that in 'this framework of revolutionary relationships, people have come to identify culture and their own liberation' (Wilson, 1988: 199). Equality for women is a recurrent theme in her story, as is the inseparability of the struggle for rights from that for cultural identity. There is an ephemerality of the built environment, necessitated by war and evident in a congress hall built to seat two thousand, with a stone podium and simultaneous translation facilities, which 'vanished almost immediately and without a trace, bulldozed to the ground' (Wilson, 1988: 202) to avoid detection by aircraft. It is in these conditions that, Wilson explains, a people's theatre has developed as one aspect of a culture of liberation. Compared to the breadth of more ordinary situations referenced in UNESCO's report on culture and diversity (see main text) the situation of camouflaged buildings and constant vigilance against attack is an anomaly. Yet through the postwar period and into the 1970s there were so many struggles for national liberation that it may seem inept to use terms such as *ordinary* in relation to conditions unaffected by them. These situations may be places in which to test retrospectively the ideas of the UNESCO report. Is culture a means of group identity? Is it potentially empowering? Looking out for occupying troops puts the problem of a possibly declining fish market in Bergen in perspective (see main text) without detracting from the importance of local economies in any affluent or majority country. But perhaps the case of Eritrea and the EPLF prompts another comparison. For the most part the culture promoted by European Capital of Culture programmes is high culture – the arts. There may be intermediate forms such as jazz and cinema, even participatory workshops and street theatre; but the presiding ethos is derived from a received supposition that culture as the arts offers universally accessible benefit. Juxtaposing this idea with the people's theatre of revolutionary situations is not a denial of that benefit, though it questions to whom it belongs, but a repositioning of it at one polarity of an axis the other polarity of which is genuinely mass culture in which production and reception converge, or, in the Forum Theatre of Augusto Boal (2000), fuse. In Forum Theatre, as Boal writes, members of the audience are able to walk on to the stage and take over a part, changing the direction of an already improvised plot. It is no longer useful to use terms like *audience* and *actor* when the roles are interchangeable. Perhaps this equates to the idea of society as a work of art proposed by Herbert Marcuse at the Dialectics of Liberation Congress at the Roundhouse in London in July 1967 as 'the most radical possibility of liberation today' (in Cooper, 1968: 185). But the Roundhouse was not camouflaged, just a redundant railway building used for arts purposes. I am left with a lingering memory from reading Amrit Wilson's text of how the cultural and the everyday merge in the culture of the EPLF, producing theatre, armed resistance and the manufacture of useful objects such as sanitary towels 'identified as a priority if women were to be effective fighters' and part of a development that has 'affected the cultural consciousness of the people as a whole' (Wilson, 1988: 201).

Development agency (Booth and Boyle, 1993: 29–30, citing Boyle, 1990; Keating, 1988; Keating and Boyle, 1986; Lever and Moore, 1986). The city's promotion was handled by the advertising firm Saatchi and Saatchi with the slogan 'Glasgow's Miles Better' (than Edinburgh).

By 1990 Glasgow had already established a significant cultural profile. John Myerscough (1988) calculated that the city's cultural economy amounted to £204 million, drawing in half a million tourists annually (cited in Booth and Boyle, 1993: 32). Garcia notes that Glasgow's year as City of Culture marked a new turn in the EC project, as the first city to be awarded the title as a means to regeneration – after Athens, Florence, Amsterdam, Berlin and Paris (Garcia, 2005: 843). She sees its legacy as 'framed in controversy' by rival claims for cultural celebrations 'in a city that has yet to resolve a remarkable divide between the wealthy and the poor' (Garcia, 2005: 845, citing Boyle and Hughes, 1991; 1994; Mooney, 2004) while accounts of the City of Culture's impact emphasise the change which occurred (and had begun) in the city's external and media perceptions, and its economic profile.

For Eisenschitz, who cites negative perceptions of Glasgow's cultural policy (above), 'Democratic cultural strategies are a response to the emergence of prestige cultural and heritage investment as cities compete for inward investment' (Eisenschitz, 1997: 170). But he admits that the gulf between dominant cultural provision and the needs of disadvantaged publics is enormous. As he continues, culture is marketed as a conflict-free solution to the ills of deindustrialisation, a mask to conceal the histories of conflict (as in labour relations) in the industries which have gone. This supports Hall's argument (1997: 218) and is reminiscent of London's Docklands redevelopment. Jon Bird writes that the erasure of histories of conflict there feeds the notion of symbolic opposition between a mythicised waterfront as 'a sinister space of Dickensian dereliction and impoverishment' and the new, 'sparkling, utopian vision of a "Metropolitan water-city of the twenty-first century"' (Bird, 1993: 123). He continues:

> In this scenario, a history of government neglect is turned into a pathology of despair, a narrative of outmoded industries and regressive labour relations, parasitical local authorities and fragmented communities, a spectre of the inner city as a virus within the social body.
>
> (Bird, 1993: 123)

The history of Docklands in the memories of local people is one of solidarity among workers resisting exploitation. Stories of strikes do not sell apartments or rent space in office towers, however, and Eisenschitz concludes:

> A democratic cultural strategy may be an apolitical irrelevancy or a challenge to the status quo. Its attractions to that status quo are not hard to see: it can be

non-confrontational and may be reached by anything from community festivals to pedestrianisation or the stimulation of the indigenous music industry.

(Eisenschitz, 1997: 171, citing Bianchini et al., 1988: 40)

The question then is whether it is democratic in determination as well as the accessibility of what it provides. Garcia concludes, as Loftman and Nevin did in Birmingham (above), that the quality of jobs produced by Glasgow's post-1990 cultural regeneration 'was often relatively poor and rarely provided the transferable skills that people need to remain in the job market' and that the impact of the cultural programme cannot be detached from that of other factors (Garcia, 2005: 861). This affirms the view that there is little evidence for the City of Culture's positive effect on local economic development while it is 'surely' an urban spectacle in which citizens 'had a good time' (Booth and Boyle, 1993: 45–46). Glasgow is cited as a model of culturally led regeneration, seen by John Urry as the British city most able to Europeanise itself through its City of Culture project (Urry, 1995: 162). The impression comes from press reports but Garcia views reliance on these as misleading: 'it is only through contrasting them with personal accounts that we can claim an understanding of cultural legacies' (Garcia, 2005: 862). There is little detailed monitoring of the EC programme.

Bergen

I move now to Bergen, in one of the richest countries in Europe. Norway's voters rejected membership of the EU but the country is part of the EC's cultural network. A publicity leaflet promoting business relocation to Bergen uses the headings of oil and gas, seafood, shipping, research and education, fjord tourism, film and music, and culture. Around 80 per cent of Norway's oil exports come from the region, and the city has the largest shipping fleet in Scandinavia and a fishing industry. Thirty thousand students (some on international programmes) use the city's University and other higher education institutions. Bergen's cultural infrastructure includes a media industry, art museums (a major collection of works by Edward Munch) and a modern concert hall built on the groundplan of a grand piano. The Hanseatic League building is a museum on the waterfront near the UNESCO World Heritage site of Bryggen, the city of wooden merchants' houses and workshops, burnt several times in its history and now restored using traditional methods and materials (fig. 6.1). The final paragraph on the promotional brochure reads 'The City of Bergen stimulates business development, and offers support in the process. Unique to Bergen are the well-established networks between industry, research and development, and the public sector' (Bergen, 2004: np).

Figure 6.1
Bergen,
Bryggen
waterfront –
UNESCO
World
Heritage site
(photo
M. Miles)

The issues here are different from those in Glasgow for obvious reasons, but the balance of local, national and international agendas remains a concern, addressed by the Commune in preparation for 2000. Bergen's year of events was, according to one of the city's public officials, one of the cheapest of the nine city projects for 2000 at around 104 million kroner, of which about 2 per cent was from the EC, 25 per cent from the Commune and 37 per cent from the national state with the rest from other public-sector sources and private-sector sponsors (Bergen, 2001: 16 – figures approximated). But it had a mandate to strengthen the economic viability of Bergen's contemporary cultural provision in the long term and overcome the city's perception as a second city (interview with William Hazell, Bergen Commune, 29 November 2004). Given a drain of cultural producers from the city in the 1990s, a key question was whether those educated in Bergen could make careers there (and not as tourist guides).

Preparations for 2000 included refurbishment of most cultural venues and some work spaces such as artists' studios. Funding was used to allow cultural producers and arts organisations to develop new projects, but also to predict future potential by monitoring its use. A maximum of one-third of an organisation's budget was provided by public investment on the ground that this could reasonably be replaced in the long term by other sources. The 2000 programme did not include many new factors. The city's main museum, for instance, had the same number of shows as usual but of a higher quality, using its own curating staff rather than inviting high-profile curators from abroad. Similarly, new music was performed by Bergen musical groups who extended their musical knowledge by working with invited composers who conducted their own works, rather than inviting celebrity conductors. There was also a balance of new media such as electronic music in alternative venues and clubs, and traditional media in venues such as the Grieghall. A key outcome of the year was an extension of transferable skills; events were opportunities to demonstrate in-house competence and there were, I was told, 'no failures in the arts sector during or post-2000' (interview with William Hazell, 29 November 2004). This follows a decision to avoid the over-rehearsed themes for the Capital of Culture year provided by Edward Grieg and Edward Munch, and to emphasise contemporary cultural production over reception. William Hazell cites audience research showing, nonetheless, that 48 per cent of Norwegians now see Bergen as a cultural centre – including 40 per cent of Oslo residents, slightly fewer of whom see their own city as one – and that the highest levels of satisfaction with the programme for 2000 were among less affluent residents with children (interview, 29 November 2004).

The place of culture in the city's self-perception seems ambiguous, still, and Siri Myrvoll writes that 'Since the main tourist attraction has always been nature, culture comes as an added extra' (Myrvoll, 1999: 44). The Capital of Culture programme was not aimed at tourism, but the city is seasonally transformed by it. Around two million people visit Bergen each year (Myrvoll, 1999: 47), mainly on cruise ships and for short stays. An annual music festival and conferences also draw in visitors. Myrvoll asserts that some of the methods used to attract tourists to the city centre are out of keeping with tradition 'or even with good taste', and that the waterside fish market – 'an embodiment of Bergen culture' – has been changed:

> Fresh fish is becoming scarce . . . the fishmongers being busy with the more profitable business of supplying smoked salmon sandwiches to the Germans or vacuum packed smoked salmon and tinned caviar to the Japanese. It has reached the point where fish vendors themselves complain: their regular customers cannot get up to the counters, they lose interest and do not return. As a result, selling fish in the fish market after the summer season is over is an unprofitable business. Bergen may be in danger of losing the foundation upon which its cultural heritage was based.
>
> (Myrvoll, 1999: 48)

My own experience on visits in November 2004 and September 2005 was that the market was not crowded with foreigners (or locals); there were several souvenir stalls as well as fish stalls, the latter selling a mix of fresh fish and tinned fish products as well as loose and pre-packed smoked salmon. I was told that fish was cheaper in shops in the city than in the market, but it was being bought in the market as well. This neither confirms nor denies Myrvoll's judgement since my visits were out of season, but her account fits with Urry's criticism that the spread of mass travel 'destroys the very places which are being visited' (Urry, 1995: 134,

Case study 6.2

From Serene Tan and Brenda S. A. Yeoh, 'Negotiating Cosmopolitanism in Singapore's Fictional Landscape'

The notion of cosmopolitanism as an important ingredient in the success of Singapore as a nation in the global arena is an issue of recent and popular interest. The 'Singapore 21 Vision' launched . . . in 1997 envisions Singapore as the ultimate cosmopolitan pit stop. This vision of the city aims to strengthen the 'heartware' of Singapore, drawing on the intangibles of society such as social cohesion, political stability and the collective will . . . The government intends to have a cosmopolitan Singapore, yet wants its citizens to feel emotionally attached to and identify with the country . . . Alongside cosmopolitan goals, the demands of nation-building are also imperative, if somewhat contradictory.

Often the discourse on cosmopolitanism is contrasted with that of 'heartlander-ism'. This cosmopolitan–heartlander divide sets better-educated, highly mobile 'cosmopolitans' against more rooted, less mobile 'heartlanders' . . . cosmopolitanism is implicitly seen to be the more exalted vision to aim for. While Singaporeans are heartlanders almost by virtue of nature and circumstance, cosmopolitans must be nurtured and developed through being exposed to the world.

The local arts scene provides a circuit of public discourse on the debates on cosmopolitanism. Literature is one of the avenues through which artistes articulate their thoughts and comments. Many have published poetry and plays, but, more popularly, there is fictional prose, through which authors fashion a landscape to reflect local society.

(Tan and Yeoh, 2006: 146–147)

Comment

Susan Buck-Morss writes of the possibility of a global public sphere in which radicalism and cosmopolitanism meet (Buck-Morss, 2002a: 10). This does not exist and is inferred only from its absence in a non-cosmopolitan radicalism and a reactionary cosmopolitanism in the War on Terror. The above description of Singapore's manufacture of a cosmopolitan cultural sphere adds a new twist to the story. It entails an immaterial production, not in cyberspace but in fiction, which stands in for a realm of cosmopolitanism which points to more than the consumption of global goods. Like the mystique of a cultural city, Singaporean cosmopolitanism is a selling point. As a cosmopolitan city, Singapore is a player in the global competition for investment while, being outside the global city loop, it wants to create a representation of itself as of equivalent status. Tan and Yeoh cite T. C. Chang: 'A relatively new area in Singapore's globalisation thrust is arts, culture, and entertainment' (Chang, 2000: 218, in Tan and Yeoh, 2006: 149), adding that the city seeks to be a performing arts hub for the region, with new urban spaces 'that will attract the cosmopolitan foreign crowd' (Tan and Yeoh, 2006: 149) such as The Esplanade – Theatres on the Bay. In the fictional texts Tan and Yeoh discuss, characters tend to be rootless. They are either heartlanders by background who remain grounded in a non-cosmopolitan world, or wander its unfamiliar spaces feeling lost. Deng, the protagonist in H. H. Tan's *Mammon Inc.* (2001) was brought up in Singapore, educated in Oxford and offered employment in New York by Mammon Inc., which has a dragon as its emblem. Deng is offered the position of adapter – who mentors executives relocating between continents in the firm's service, global nomads like herself. Deng is almost at home in New York, a corporate jet-setter and cosmopolitan whose attitudes are open and who feels at ease in the non-places which characterise a post-industrial society (Augé, 1995) – as I am at home in discourse. Marc Augé writes:

> The link between individuals and their surroundings in the space of non-place
> is established through the mediation of words, or even texts. We know . . .
> that there are words that make image – or rather images: the imagination of
> a person who has never been to Tahiti or Marakesh takes flight the moment
> these names are read or heard . . . Certain places exist only through the words
> that evoke them . . . banal utopias, clichés.
>
> (Augé, 1995: 95)

Thinking back to Walter Benjamin's discovery of curiosities evoking a latent utopia in the arcades of Paris (Chapter 3) I wonder if behind the banality of utopias evoked in words there is buried nonetheless a mark of another utopia which might, in utterly different conditions, evolve towards realisation and is partially, obliquely glimpsed in the fictional cosmopolitanism of elites in Singapore.

citing Mishan, 1969: 142). It can be argued that tourism is a democratisation of travel, or that programmes designed to celebrate a city's cultural histories and vitality may adapt both according to a perceived European cultural identity and the requirements of competition in global markets. Local cultures (ways of life) may be overlooked as conditions are shifted to those thought most favourable for cultural tourism, its horizons coloured by histories of the exotic and the wishful, and attraction of inward investment. Culture tends to displace cultures in such scenarios, while high culture displaces popular culture. This need not be the case: it is a matter of policy, planning and participation. Perhaps it is a matter of managing myth as well. For Frank Eckardt, 'face-to-face communication between intellectuals is the root of the Weimar myth' (Eckardt, 2006: 122). The city's tourism rests on visitors' desires to experience sites associated with Goethe and Schiller, Liszt or Nietzsche, all of whom resided in Weimar, their houses now preserved as museums. After German reunification Weimar's population declined along with that of other cities in the former East, to be stabilised by a manipulation of city boundaries, suburbanisation and encouragement of a cultural and touristic identity. Weimar was City of Culture in 1999 despite facing economic difficulties, and Eckardt sees this as a turning point in perceptions of the city by its residents despite the fact that 'The mapping of Weimar is dominated by the attractions of the city which are upheld by the external world' (Eckardt, 2006: 129). In the 1920s the Bauhaus, in Weimar when it was identified with a fledgling German republic, sought to change the map.

Looking ahead – cultural diversity?

Global competition encourages the production of blockbuster art exhibitions and flagship buildings which have become motifs of an aestheticisation of urban life and space. Greg Richards argues that spectacles are important in the marketing strategies of cultural tourism and that, as a consequence, 'people in the cultural sector are being pulled into the tourism arena, whether they like it or not' (Richards, 1999: 27). This leads to resistance. Bianchini draws attention to another problem:

> The participation of disadvantaged ethnic and racial communities in public and cultural life in European cities is particularly problematic. Many working people belonging to Asian, African, Afro-Caribbean, Turkish and more recently East European ethnic communities are trapped in the unskilled sectors . . . They are frequently discriminated against in housing, education, policing and the provision of other urban services.
>
> (Bianchini, 1993: 205)

He argues that cultural policy can be used to address this but needs to deal with its depoliticisation since the 1980s, and adopt an emphasis on localised, grass-roots engagement. I agree and disagree: I agree because cultural provision has a potential for democratisation and intercultural collaboration, which may inform other areas of social formation and power relations; I disagree because the expediency of culture (Yúdice, 2003) tends to undermine this while it is no substitute for housing, health or education. Evidence cited from Glasgow and Birmingham (above) implies that culturally led redevelopment offers little stable or skilful employment. Nonetheless, the model of grass-roots-led development would be interesting, and raises the possibility of cultural work as having a role not unlike that of development work in the non-affluent world when NGOs attempt to hand over control of projects and agendas to local people.

Drawing the chapter to a close I look now at UNESCO's report 'Our Creative Diversity' (1996), using the summary version. Under the heading 'Creativity in politics and governance', the summary report states:

> The nurturing of collective creativity also means finding ways of help-ing people to create new and better ways of living and working together. Our social and political imagination has not kept pace with our scientific and technological imagination. It has been said that central government, which has usurped more and more power to itself . . . There is a need to better explore methods and procedures such as delegation of authority and decentralization in ways that promote access to voice and to power.
>
> (UNESCO, 1996: 24)

The report is bland but uses terms – political imagination, decentralisation, voice, power – which imply a position framed by the needs and aspirations of the majority (non-affluent) world. Among other insights the report states a definition of culture as 'ways of living together' (UNESCO, 1996: 14); and development as economic growth to which culture is purely instrumental, or as a broader ambition for 'freedom to live the way we value' (UNESCO, 1996: 14). It refuses a reduction of culture to the promotion of economic growth, and continues:

> There is, in addition, the role of culture as a desirable end in itself, as giving meaning to our existence. This dual role of culture applies . . . in relation to other objectives, such as sustaining the physical environment, preserving family values, protecting civil institutions in a society, and so on.
>
> (UNESCO, 1996: 14)

I agree and disagree again, and will try to explain why. I am encouraged by the report's refusal of cultural expedience and assertion of cultural values. I agree that culture (including but not only the arts) is one of a number of ways in which meanings are produced and the world made intelligible. I engage with visual art, literature and music as expressions of specific, historically produced responses to conditions, which help me articulate my own responses to the conditions in which I live. I am also sympathetic to the idea that the arts can assist in disseminating critical views of environmental destructiveness. But I wonder if the report substitutes one kind of expediency for another in seeing culture as able to achieve goals it supports but rejecting its role in relation to goals it refuses. I am unsure that the traditional values it affirms are compatible with democratic social forms or equality of opportunity, for instance. The report is vague about which civil institutions should be protected, using the phrase 'and so on', and I

Case study 6.3

Daniela Brasil, *Folge mir (Follow me)*, Weimar, 2004

Figure 6.2 *Daniela Brasil,* Folge mir, *spoof traffic sign, Weimar, 2004 (photo D. Brasil, reproduced by permission of the artist)*

Comment

Folge mir comments on modes of behaviour characteristic of German city life by intervening temporarily and unexpectedly in the context of the existing system of road traffic signs. The intervention is unexpected because the traffic system is highly regulated, each sign corresponding to a perceived necessity so that a sign exhorting citizens to kiss is outside its universe. Brasil writes: 'As a discrete provocation, the project tries to seduce the passers-by, instigating a more affectionate behaviour in the city's public spaces. It irreverently and silently offers a sentimental dialogue attempting to extract people from their habitual selves, and the city from its routine' (email to author, 9 January 2006). At a seminar presentation of the project Brasil mentioned that in Brazil, her home country, or in Portugal where she worked for several years before moving to Weimar in 2003, it would be ineffective because drivers pay scant attention to traffic signs. In Germany they obey them. This is a caricature, too much like the German revolutionaries queuing to buy platform tickets before taking over the railway station, yet has a resonance. The habits of road use are as worthy of investigation as those of cuisine or interior decoration, indicate values embedded in a social formation at a specific time and carry the baggage of received conventions of behaviour in a specific society. Most traffic signs say things like 'No Left Turn' or 'Stop'. This one says 'Kiss me'. Should people turn to the person nearest to them at the time to make the gesture of affection, or peace as in the Catholic mass, except that Weimar is a Lutheran city? Thirteen signs were installed in the centre of Weimar in 2004 using similarly imperative but not identical texts, and remained on site until they were stolen or damaged, 'Kiss me' being attacked more forcefully than others. The project had official permission, but the sign was knocked over and bent. This surprised the artist but started a public discussion through the local radio stations 'about the "Weimar sex-appeal" and the different habits of kissing, as affection or greeting, in latin and anglo-saxon countries' (email to author, 9 January 2006). An article on the City of Culture project in *German News, The Magazine* (August–September 1998) says 'Weimar plans to present itself in a playful way within a European framework. There will be a workshop for artistic potential . . . living, playing and listening' (www.germanembassyindia.org/news/april97/10gn04.htm).

wonder if non-institutional and self-organised aspects of a culture are included. At the same time, I prefer civil society to military rule, obviously; and am excited by the prospect of 'cultural freedom' offered at the end of the paragraph quoted in part above (UNESCO, 1996: 15). There is much I read as critical and useful in the report. For example: 'One of the most basic freedoms is to define our own basic needs' (UNESCO, 1996: 15). It affirms pluralism in culture and governance, advocating respect for minority groups and defining culture as broader than the arts: 'The arts . . . grow out of the soil provided by the more modest routines of daily life' (UNESCO, 1996: 23). The report argues for new cultural policies for diversity, and support for initiatives which are autonomous of national agendas. It states: 'Consolidating social integration with respect to ethnic and cultural diversity . . . is a major public policy challenge' while support for experimental art is 'an investment in human development' (UNESCO, 1996: 41).

Conclusion

Common to UNESCO's aspirations for cultural diversity and empowerment, and the EC's Capitals of Culture programme, is a reassertion of local and regional cultures in face of globalisation and creeping Disneyfication. Urban cultures are privileged in the EC programme for an industrialised, heavily urbanised continent, but it lacks an awareness that the vitality of urban cultures is more likely in an antagonistic democracy than in pursuit of homogeneity or the bland concepts the EU tends to advance as potentially inclusive. I borrow the concept of antagonistic politics here from Leonie Sandercock's discussion of mongrel cities (2006), and put it in context of writing by Iris Marion Young (1990; 2000) and Susan Buck-Morss (2002a). I conclude by quoting Sandercock:

> In demographically multicultural societies and politics, conflicts over values and lifestyles, ways of being and ways of knowing, are unavoidable. As long as there is global movement of peoples and their accompanying cultural baggage, consensus will only ever be temporary . . . An agonistic politics entails broad social participation in the never completed process of making meanings and creating values. An agonistic politics implies 'the end of mainstream' in terms of the end of a single dominant culture in any polity, perpetual contestation over what is or might become common ground, and negotiation towards a sense of shared destiny.
>
> (Sandercock, 2006: 47–48)

Cultural celebration is one of many vehicles for this if released from the obligations of economic growth, and illusion of humanity as a happy family like those who used to appear in television advertising for washing powder or consumer durables in the 1960s.

Notes for further reading

Chapters 5 and 6

Sharon Zukin's *The Cultures of Cities* (1995) is relevant here as to Chapters 3 and 4. Of works dealing with the specifics, however, of the cultural sector Allen Scott's *The Cultural Economy of Cities* (2000) is the most detailed and could be read with Mike Featherstone's *Undoing Culture: Globalization, Postmodernism and Identity* (1995). On cultural planning Graeme Evans contributes a large volume of data in *Cultural Planning: An Urban Renaissance?* (2001), as in several papers listed in its bibliography. Given the appeal to dominant cultural frameworks of much cultural policy, it may be interesting to compare the general arguments and assumptions of such books with George Yúdice's *The Expediency of Culture* (2003), which makes a case against utilitarianism in the arts and presents material on cultural development work in Latin America. And while policy agendas categorise groups as excluded, *Spaces of Social Exclusion* by Jamie Gough, Aram Eisenschitz and Andrew McCulloch (2006) gives a more direct view of the situation of marginalised publics.

Adorno's essays on the culture industry are reasonably accessible and collated in *The Culture Industry: Selected Essays on Mass Culture* (1991), which could be read with *The Stars Down to Earth and Other Essays on the Irrational in Culture* (1994) which includes his study of a newspaper horoscope column from Los Angeles. Papers (mainly from leftist journals in the 1930s) on Expressionism and socialist realism by Adorno, Benjamin, Bloch and Lukács are collated in *Aesthetics and Politics* (Adorno et al., 1980). My own experience has been that trying to understand Adorno via interpretations of his writing is often less helpful than reading it in the original, with the exception of Robert Witkin's *Adorno on Popular Culture* (2003), which is particularly clear on film and popular music. Apart from that, Herbert Marcuse's *The Aesthetic Dimension* (1978) gives a short argument similar in some ways (not all) to that of Adorno's *Aesthetic Theory*, and Marcuse's collected works (1997) now appearing in a series edited by Douglas Kellner (Marcuse, 1998; 2001; 2005), include overlooked but potentially key contributions such as his essay on French literature under the German occupation in the 1940s. Walter Benjamin's work is covered in the notes to Chapters 3 and 4.

The everyday aspect of urban culture is less documented than the spectacular, but encountered in Ruth Finnegan's *Tales of the City* (1998). Ben Highmore's *Everyday Life and Cultural Theory* (2002a) is an excellent introduction to a complex field, to which his edited *Everyday Life Reader* (2002b) adds a range of texts from Sigmund Freud, Michel de Certeau, Leon Trotsky, Trinh T. Minh-ha and Stuart Hall among others. These could be read with Michael Gardiner's *Critiques of Everyday Life* (2000). The relation of visual and aural cultures to the formation of urban bohemias is covered admirably in Richard Lloyd's *Neo-Bohemia* (2006), which adds a North American view to that of Elizabeth Wilson's *Bohemians: The Glamorous Outcasts* (2000). In a political terrain Margaret Kohn (2003) investigates the House of the People as a site of working-class cultural and political formation in Europe, with considerable insight into the concept of a public sphere. For discussion beyond a Western focus, *Culture and Global Change*, edited by Tracey Skelton and Tim Allen (1999), has sections on culture and development, and culture and human rights; and *Liberation Ecologies*, edited by Richard Peet and Michael Watts (1996), gives a range of incisive commentaries on emerging socio-political, cultural and ecological perspectives from the dominated world. The cities included in the EU Capitals of Culture programme are best accessed via their websites (and many others) using a search engine. But for a cultural history of Weimar prior to its reinvention as a City of Culture I suggest John Willett's *The New Sobriety* (1978), and a chapter by Frank Eckardt (cited above) in *Small Cities*, edited by David Bell and Mark Jayne (2006 – cited in the notes for Chapters 1 and 2). Dealing with urbanism and planning, and the development of multiple modernist city forms, Helen Meller's *European Cities 1890s–1930s* (2001) includes chapters of the transition to modern forms of Barcelona, Birmingham, Blackpool, Budapest, Hamburg, Lyons, Marseilles, Munich, Nice, Vienna and Zlín. The last is particular interesting, a modernist equivalent of a garden city founded by Czech shoe manufacturer Tomás Bat'a in the 1920s. Cities including Berlin and Bucharest (cited in Chapter 7) are covered in terms of cultural agendas and architecture in times of change in *Architecture and Revolution*, edited by Neil Leach (1999). Conventions of cultural identity are challenged in Etienne Balibar's essay 'Algeria, France: One Nation or Two?' in *Giving Ground*, edited by Joan Copjec and Michael Sorkin (1999), which includes Dean MacCannell's essay on Celebration, Florida (cited in Chapter 5).

Part Four

Interventions

7 Cultural cross-currents

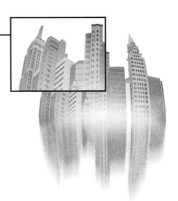

Learning objectives

- To examine intersections of urban cultural currents in late modern times
- To look outside Western perspectives of urban development in a period of globalisation
- To understand the role of public monuments in times of change

Introduction

In Chapters 5 and 6 I investigated the culture industries and the European Union's Cities of Culture programme, ending by asking what potential existed for cultural work as a means of empowerment. In the next chapter I resume that discussion through a specific case of an arts project in a social housing district of Lisbon. In this chapter I look at the cultural cross-currents produced after the dismantling of the Berlin Wall in 1989 and disintegration of the Soviet Union in 1991, and their manifestations in former East bloc cities. I look at the current situation in Yerevan, Armenia, to identify such currents; and at the role of public monuments when power shifts through the case of Grūtas Park (Stalin World) in a forest in Lithuania, to which more than seventy Soviet-era statues were removed in 1999. Being unsure how to read these displaced monuments I compare the positions of Renata Salecl (1999) and Laura Mulvey (1999). Because a context for the shifts of power discussed is the globalisation of capital, communications and culture – reflected in a move towards consumerism which fuelled the collapse of state socialism in the East bloc – I begin by outlining aspects of globalisation relevant to the chapter (though economic globalisation as such is outside my scope). In case studies I quote from a novel by French economist Jeanne Hyvrard (1975);

and Jyoti Hosagrahar's account of the division of Delhi into European and Indian districts after the 1857 insurgency. Finally, I comment on a photograph of an experimental settlement, Arcosanti in Arizona.

Global contradictions

The global and the local

I begin with three snapshots of a globalised world. First, Arjun Appadurai writes that policies on copyright, bio-patent, environment and trade in the affluent world 'set the stage for life-and-death decisions for ordinary farmers, vendors, slum-dwellers, merchants, and urban populations' (Appadurai, 2001: 2). Second, Zygmunt Bauman (1998) sees globalisation as producing increased mobility for the rich and control of movement for the poor, classed as migrants and asylum seekers, while local initiatives to regulate resource extraction and environmental destruction are disabled by transnational agreements deregulating trade and labour. Third, David Camacho notes that a concern for human welfare ecology is brought into sharp focus by nuclear plant accidents, oil spills and global warming, adding that environmentalism 'has tended to exclude substantive participation by people of color, although people of color have organized around environmental issues at an unprecedented rate since the late 1980s' (Camacho, 1998: 11). A picture of a world dominated by transnational companies operating with larger economies than those of medium-sized nation-states, crushing local resistance to resource exploitation and habitat destruction, is familiar. Similarly, there is a broad acceptance that consumerism serves the requirements of such corporate interests, and is enforced through branding and globalised media of news, communication and entertainment. If this is regarded as unsustainable, as the ecological footprints of Western societies far outstrip the planet's finite and non-renewable resources of energy and raw materials for industry, opposition within affluent societies comes mainly from elite groups. As contributors to Camacho's *Environmental Injustices, Political Struggles* (1998 – see also Guha and Martinez-Alier, 1997) show, the import of another kind of elitism in expertise and debt-producing technologies is inappropriate to the environmental and social problems faced in the majority (non-affluent) world.

But this is not the only picture. Demonstrating that power produces resistance, dissent is now as global as the operations of capital. Appadurai observes social forums emerging

> to contest, interrogate, and reverse these developments and to create forms of knowledge transfer and social mobilization that proceed independently of the actions of corporate capital and the nation-state system . . . [which] rely on

strategies, visions, and horizons for globalization on behalf of the poor that can be characterised as 'grassroots globalization.'

(Appadurai, 2001: 3)

Examples include the World Social Forum (Houtart and Polet, 2001), and the succession of anti-capitalist insurgencies which began in Seattle in December 1999 (Starhawk, 2002; George et al., 2001). In French the term *mondialisation* helpfully differentiates these currents from the anti-democratic structures of globalised capital. As Raff Carmen writes from experiences in Zambia and Burkino Faso, 'The situation would be hopeless if indeed the culture of power were hegemonic but . . . there always will be the countervailing power of culture. This culture has the power to resist' (Carmen, 1996: 157). François Dufour writes of French farmers' action against junk food: 'In addition to the support we'd expected from activists, there were many who rallied behinds us from unexpected quarters, including lemonade merchants who refused to sell Coca-Cola' (Bové and Dufour, 2001: 30), and a key characteristic of the emerging political map after globalisation is the prevalence of coalitions of disparate elements around single issues affecting lifestyles as well as habitats. Local groups network globally, and the growth of fair trade consumption in the West demonstrates that consumers have some power to inflect if not to drive processes of environmental justice via the mechanisms of global trade.

The localised self-organisation of the poor has led to the legalisation of informal settlements (Fernandes and Varley, 1998), claims to women's rights (Kabeer, 1994) and advances in participatory means to shape a social world (Berg-Schlosser and Kersting, 2003; Guha, 1989). But none of the currents flows in one direction only. For instance, Brenda Yeoh observes a move from transnational to local cultural scale in the promotion of cities (Yeoh, 2005: 945), citing John Clammer:

> It is the cultures of urban spaces that are most immediately and directly influenced by globalisation (in terms of consumption patterns and tastes, fashion, architecture, media and new forms of material culture) and equally it is urban cultures (intellectual trends, economic and technological innovations and again the media) which largely constitute so-called globalisation.
>
> (Clammer, 2003: 403–404, cited in Yeoh, 2005: 945)

Yeoh emphasises the roles of cities as political-economic entities decreasingly constrained by national borders as players in transnational regional economies and political formations. The Pacific Rim is such a case, linking China and Japan with Malaysia, Thailand, Indonesia and Australia while housing diverse cultures. Yeoh adds, 'Cultural imagineerring is sometimes a strategy of reconstruction occasioned by regime change' as in Jakarta in 1998 (Yeoh, 2005: 947), implying the use of cultural means to promote a reformation of socio-political structures in the manner of city rebranding.

I wonder if the separation of culture as the arts and media, and cultures as ways of life, which I adopted in Chapter 2 is viable when a consumerist way of life is both engine and outcome of an economic domination propelled by mass-media communication. In a suitably postmodern way, the categories of art and everyday life seem no more serviceable than base and superstructure, or cause and effect, when the interactions of factors within any set of conditions are intricate and complex (Cilliers, 1998). Peter Worsley (1999a) proposes an extended definition of culture to encompass subcultures, and notes Antonio Gramsci's argument that hegemonic cultural means to economic and political conformity are resisted in counter-cultural forms. He argues that culture is too often seen as either a discrete sphere or a residual category, and that 'we need to see culture as a dimension of all social action, including economic and political' (Worsley, 1999a: 21).

The Eastern and the Western in Europe

Issues remain of resistance but also of slow, incremental shifts and mutual permeations. For instance, the end of the Cold War appeared sudden in iconic images of the dismantling of the Berlin Wall in 1989, flashed around television screens worldwide as it took place. Yet the prelude was a slow move towards consumerism in the East bloc from the Krushchev era onwards, accompanied by periodic if irregular relaxations of cultural policy. There clearly were differences between East and West, and these were evident in daily life, but they were not absolute. Judd Stitziel describes the growth of a fashion industry within the German Democratic Republic (GDR), for example, and a shift from utility to a design-based approach: 'The production of expensive, desirable clothing and its sale in special stores also represented a valuable political and cultural asset for a regime that promoted "clothing culture", promised to fulfil its citizens' increasing needs, and competed with the West on the basis of individual consumption' (Stitziel, 2005: 119). A factor in the unrest immediately prior to the end of the Wall was an economic decline and hence reduction in availability of consumer goods in the GDR, once appetites for consumption had been enlarged by its own production and distribution systems – if always with less glamour and efficiency than in the West. Bureaucrats asking sales staff in Leipzig in 1965 why they failed to argue with customers who complained about GDR textiles 'received a blunt response' that the customers were right (Stitziel, 2005: 163). Similarly, opposition to the regime did not arise suddenly. Stephen Barber writes of Karl-Marx-Allee in Berlin, called Stalinallee when it was built in the 1950s, that construction workers had rioted against the provision of luxury apartments for party bureaucrats there. In 2001 he recalls that in the previous decade (after reunification), 'the exterior walls . . . had attracted an intricate layering of graffiti that had metamorphosed in inflection from year to year, fluidly crossing sexual, political and emotional ground. Now this graffiti had been obliterated' (Barber, 2001: 64). The site of the Wall is now even more transformed in recent construction, as around Potzdamer Platz (Siegert and Stern, 2002).

Boris Kagarlitsky writes that in the decade after 1989 Eastern Europe has become 'part of the world capitalist economy – its new periphery' (Kagarlitsky, 2001: 53). He points to contradictions between the beginnings of East-bloc consumerism and a failure to deliver its promises:

> the road of development that had been chosen ensured that the countries of Eastern Europe would increasingly be drawn into the world economy as the periphery of the West. Their dependency increased steadily throughout the 1970s and 1980s. At the same time, their internal problems were not being solved. From the point where the system proved incapable of satisfying the consumer expectations which it was itself calling forth, it ran up against growing political disaffection, multiplied by narrow-minded resentment. The movement of 1989 was just as much an uprising by enraged consumers as it was the revolt of an awakened 'civil society'.
>
> (Kagarlitsky, 2001: 55)

After 1989 Berlin saw a building boom, as documented in Hubertuius Siegert's film *Berlin Babylon* (2001). But construction entailed demolition, and uncovered sites of the Nazi period while clearing those of state socialism. As Ralph Stern observes, the film 'intersects a myriad of discourses. Some of them are specifically connected to the history and politics of Berlin as well as the ideological inclinations of many of Berlin's contemporary architectural and political figures' (Siegert and Stern, 2002: 118). He continues that other readings concern public memories and the city's contested identity in a period of highly visible globalisation. Meanwhile a walk through Berlin connects monuments from the Whilhelmine era prior to 1918 to the Television Tower erected in the Soviet period to be seen from the West, a Holocaust Museum and the brash constructions of a market economy. The traces of conflicting and disputed histories are everywhere, and questions arise as to how such traces are interpreted, whether they should be buried or displayed, whether a society's past sign systems should be replaced, displaced or disgraced. I now consider such issues in terms of the changing appearance of Yerevan, and the re-siting of Soviet era statues in Lithuania, in context of the roles of monuments in social formation.

Displaced narratives

Monuments and power

As iconic images, public monuments convey to a society's members the values they are required to emulate in durable materials such as stone or bronze from which the inference is that the narratives will endure as long as the materials. Generals, colonial adventurers, politicians and merchants stand on plinths so that spectators look up to them, and if the individuals depicted are forgotten the form of the monument continues to impress. The use of naturalism in the nineteenth century does not detract from the separateness of monuments and those whom they depict – traditionally a ruling class – from their intended publics. A tendency to represent authors and film stars at street level is a democratisation of the form of the monument, and access to the category of representation when fame replaces wealth or power, but retains a distinction between the star and a mass public who might hope the stardust rubs off when they are photographed next to the street-level bronze of Cary Grant or James Joyce. Although monuments provide recognisable forms by which to navigate a city, and provide tourists with photographs imitating those of the brochure (Urry, 1990: 138–140), they are recontextualised as accessories for a city's marketable image in another assertion of a determination of a dominant image. The monumental aspect of a city, that is, far from housing a democratic public sphere constitutes a public realm which is exclusionary for many citizens. I have written elsewhere on attempts by contemporary artists to democratise or subvert the monument (Miles, 2004: 98–106) and am concerned here with how readings of monuments change in moments of revolt, or are co-opted to other agendas in less challenging times.

Joanne Sharp, Venda Pollock and Ronan Paddison interpret Antony Gormley's *Angel of the North* as denoting a depoliticised reworking of a city's skyline: 'public participation is not high on the agenda in this form of "authoritarian populism" . . . In claiming to be a signifier for the city as a whole . . . it hides the inclusions and exclusions inherent in *any* singular vision for a community' (Sharp, Pollock and Paddison, 2005: 1014). The same could be said of Nelson's Column, the Statue of Liberty or the Eiffel Tower (Warner, 1985: 6). In times of revolt attention attaches to monuments as visible signs of a regime so that their destruction stands for the end of a history or trajectory. An eye-witness to the fall of the Vendôme Column in the Paris Commune of 1871, for instance, recalled, 'The column lies on the ground, split open . . . Caesar is lying prostrate and headless. The laurel wreathed head has rolled like a pumpkin into the gutter' (from Edwards, 1973: 147–148, cited in Wood, 1999: 119). A photograph taken the next day shows the head attached and perhaps the event was aligned by the witness to a fabled history in which the heads of tyrants roll into ditches. When the bronze representation falls, onlookers seek to participate in the downfall by giving such accounts, and by re-enacting the shift of power in violent attacks on the statue. An obvious explanation for this would be that such acts would be suppressed had the shift of power not taken place, and are its living proof. Another is that people feel a need to prove their own presence in historical events, as by taking away bits of the Berlin Wall when it was demolished. Sergusz Michalski notes that the statue of Stalin at Felvonulási Square in Budapest (Mikus Sándor, 1951), made from melted-down bronze from bourgeois-era statues, was destroyed in the Uprising of October 1956, accompanied by 'especially vivid acts of public denigration' (Michalski, 1998: 141). He includes a photograph showing the head being hit with a pole and being dragged by chains across tramlines (Michalski, 1998: 143, fig. 96). Reuben Fowkes observes that the statue was the focus of public rituals in the early 1950s: 'There are echoes of the practices of the pre-modern era and even of paganism. The *Stalin Statue* resurrected the ancient habit of venerating images as if they were identical with the person they represented' (Fowkes, 2002: 81). He notes that the destruction of the statue followed Khrushchev's denunciation of Stalin, and a Central Committee decree in June 1956 'On Overcoming the Cult of Personality and its Consequences' (Brown, 1998: 305, cited in Fowkes, 2002: 81). The crowd pulled down the statue using wires, blowtorches and trucks in 'a ritualized act of crowd justice that marked the crossing of a revolutionary threshold' (Fowkes, 2002: 82). Onlooker István Kállay stated that when it fell, 'it was like a rubber ball, it jumped up a couple of times before coming to rest' (interview, Oral History Archive vol. 282, p. 23, cited in Fowkes, 2002: 82).

The re-enactment of change appears to complement what Joe Kerr describes as a need for material presences to commemorate significant public events: 'the act of abstracting through the erection of monuments that permanently inscribes the image of that event' (Kerr, 2001: 71). Kerr cites The Monument in London (to the fire of 1666) as one example of 'the tangible trace of collective memory' (Kerr, 2001: 72), and the Lenin Memorial, designed by Berthold Lubetkin for a site in London to celebrate Anglo-Soviet co-operation in the war years (destroyed on a cooling of relations in 1948), as another. The process Kerr describes took place particularly in Russia after 1917. Lenin's idea, informed by Tomasso Campanella's *City of the Sun* (published in translation by the Petrograd Soviet in 1918 – Mulvey, 1999: 222), as reported by Anatolii Lunacharskii (head of the People's Commissariat of Enlightenment), was that public spaces should be embellished with monuments to the achievement of socialism. As Susan Buck-Morss writes, 'The masses would *see* history as they moved through the city' (Buck-Morss, 2002b: 42). Lenin's programme for public monuments and street theatre was effected in the weeks immediately following the October Revolution, as he counted the days until the Revolution had lasted longer than the Paris Commune. Materials were in short supply and an obelisk was recoded to provide a new monument by erasing the names of Tsars and inscribing those of revolutionary forerunners: Marx, Engels, Liebknecht, Lasalle, Bebel, Campanella, Meslier, Winstanley, More, Saint-Simon, Waillant, Fourier, Jaurès, Proudhon, Bakhunin, Chernyshevski, Lavrov, Mikhailovskii and Plekhanov (Buck-Morss, 2002b: 43, fig. 2.2; Mulvey, 1999: 223). It may be that erasures of a history by destruction of its public monuments are necessary. If the people were to see history as they moved through the city they were to see its redirection in such erasures, as well as its celebration in street theatre and agit-prop performances.

Yerevan

A question arising from this material is whether the form of the monument can be detached from the specific history it relates. An obelisk is perhaps sufficiently abstract a form to allow reinscription, a *tabula rasa* distanced almost beyond connection from its origin as a sacred monument in archaic Egypt. In other cases, notably the form of the bronze figure on a plinth, old readings may simply be adapted for a new cast of heroes while the need for the heroic as a category is unquestioned. To give an example of what I mean, in Yerevan, Armenia (a constituent republic of the Soviet Union from 1936 to 1991, after being part of the Soviet Transcaucasian Federation since 1923), a 21-metre statue, *Mother Armenia*, on a 34-metre plinth, was erected on the site of a 16.5-metre statue of Stalin by sculptor Sergei Merkurov which had occupied the site from 1950. Across the city from Stalin had been a statue of Lenin by the same sculptor, removed in 1991. A guidebook cites a Soviet writer in 1952:

> Stalin [is depicted] in a characteristic pose of dynamic movement, supreme composure and confidence . . . one hand in his coat-breast and the other slightly lowered as though his arm, swung in rhythm with his step, has for one brief moment become free in space. Stalin's gaze rests upon the splendour of the new Armenian socialist capital, upon its new handsome buildings, its wide green avenues, upon the central square in the opposite end of the town. There, Lenin, in his ordinary workaday suit, has swung abruptly around in that characteristic, impetuous sweeping way of his, so dear and familiar to every Soviet man, woman and child.
>
> (source unstated, cited in Holding, 2003: 89)

If they were known in this way it was a fulfilling of Lenin's programme that the mass public should see their history around them. It may be that *Mother Armenia* – which was pointed out to me in a positive way by several people I met in Yerevan in 2005 and 2006 – meets a similar need in context of national revival today, retaining the principle that a national identity is figured in a city's monuments.

Mother Armenia stands in a memorial zone at the top of a steep slope partially developed in Soviet times as flights of steps, gardens and fountains called The Cascades. A Museum of Contemporary Art is under construction on the site to house a collection of Western, modernist art. Currents of residual and post-socialism, nationalism, internationalism and a memory of genocide in 1915 cross in Yerevan today as the city centre is rebuilt in redevelopment schemes financed by members of the Armenian diaspora. The central district is a mass of deep holes in the ground and construction sites from which standard steel-and-glass towers begin to rise (in April 2006). An incipient, aspirant world city is surrounded by nineteenth-century buildings in local pink and grey stone in a city the plan of which (by Alexander Tamanian, 1926) merges aspects of the English garden city with Campanella's concentric circles. Several of the older buildings have had their stones numbered (fig. 7.1) in preparation for removal to storage and eventual reconstruction on another site, though there is doubt as to whether this will happen and what effect it would have, possibly creating an *ersatz* heritage mini-city while the city centre mimics global city forms in a country desperately short of resources by Western standards. I saw this building in October 2005. By April 2006 it had gone.

The old monuments have been replaced, Lenin by a video screen in a flower bed in Republic Square. The screen broadcasts weddings, current events such as the procession to the national memorial on Genocide Day and advertising. Other monuments have been replaced by figures of national history, in several cases in more or less socialist realist style. But there is an alternative representation of a transitional society in new cultural tendencies, mainly ephemeral, working in video or performance and taking as a context that there is no clear context any more – while an older generation were able to establish themselves in context of Perestroika.

Armenia was a cultural hub in Soviet times, its National Gallery containing important works of Russian and international modernism, and the Perestroika generation could inherit that legacy. For emerging artists today the context is confusing – the Soviet past, and behind that national culture; international modernism, which they know to be already over while they read its texts for the first time; postmodern irony, gesture and codification; the 1915 genocide and the ever-present iconic image of Mount Ararat (now in Turkey). The Armenian critic Vardan Azatyan writes that in the 1980s 'Armenia was experiencing the death of the whole state system which, like a birthday, was indeed a matter of celebration'

Figure 7.1 *Yerevan, building on Republic Street prepared for removal (photo M. Miles)*

but that after a war with Azerbaijan which resulted in the termination of energy supplies, and a devastating earthquake in 1988, 'It seemed that death was literally walking the streets of the city' (Azatyan, 2006: 8). This morbidity was adopted (a few months prior to the earthquake) by an experimental artists' group called 3rd Floor whose members staged a happening – *Hail to Artists' Union from the Nether World: Official Art Has Died* – in which they walked through an exhibition 'like the characters from some phantasmagorical play', heroes from the abyss moving silently and spectrally as they viewed the works and left the room (Azatyan, 2006: 8).

Some of these dissident artists were accommodated in the state system so that 'the ghosts in the exhibition . . . were legitimate parts of the official institution they wanted to sabotage' (Azatyan, 2006: 8). This replicated Soviet-era partial

Figure 7.2 CCCP – restaurant sign, Yerevan, 2005 (photo M. Miles)

accommodation of dissident culture, leading to an aspiration – largely a fantasy – on the part of such licensed dissenters that they could expand the vocabulary of legitimated cultural signs. But the difficulty after 1991 and the break-up of the Soviet Union is that the structure of state system or dissidence – both other to Western structures – gives way to one in which the immediate possibility for non-established artists is to seek accommodation within the codes of Western art. At the same time, postmodern irony has entered the cycle of wealth creation, in a rusting steel hammer and sickle advertising a restaurant called CCCP (fig. 7.2), near the city centre and adjacent to the offices of Porsche.

The redundant statues

An immediate problem for civic authorities in Eastern Europe after the dismantling of the Berlin Wall in 1989, and in Russia and other constituent republics of the former Soviet Union after 1991, was what to do with the statues of the old regime. Laura Mulvey visited Russia in 1991 to make the film *Disgraced Monuments* (1992) with Mark Lewis. She notes Walter Benjamin's observation in *Moscow Diary* of a shop selling Lenins in all sizes; and adds from her own experience:

> The studios of the official artists we visited . . . were filled with maquettes of Lenin statues dating from the 1950s and 1960s. The poses had become fixed and stereotyped: Lenin with one arm outstretched, with both arms outstretched, standing still, walking forward, sometimes holding a cap, sometimes wearing a cap and so on. One favourite anecdote was of a statue which had got muddled and appeared with Lenin both holding and wearing a cap.
>
> (Mulvey, 1999: 222)

Mulvey continues that the problem of what to do with the monuments of past regimes is a problem of historical memory: 'If seventy years of communist rule had been marked by these waves of iconoclasm, many of the people we interviewed in 1991 said that an ability to live with monuments . . . would now mark an ability to live with the past' (Mulvey, 1999: 222). Susan Buck-Morss argues that, if revolutions are legitimated by the histories they appropriate, 'The suturing of history's narrative discourse transforms the violent rupture of the present into a continuity of meaning' (Buck-Morss, 2002b: 43). And Mulvey reads the growth of a Lenin cult after his death in 1924, when his body was embalmed and his portraits exhibited like icons, as a re-emergence of belief 'starved of objects to venerate' since iconoclasm of the October Revolution (Mulvey, 1999: 222).

When their ideological currency became worthless, Soviet-era statues were broken up, though in Moscow some were resited temporarily to an open-air museum. In Poland bronzes were returned to a foundry at Gliwice to be melted down but the price of bronze was too low so the factory kept them for a museum. In Budapest a site was allocated for Statue Park (Szóbórpark), designed by Akos Eleod, where, beyond an ornate gateway, Lenin and Marx now address tourists, surrounded by hoardings and electricity pylons on the outskirts of the city, and a souvenir shop sells tins of The Last Breath of Communism (Rugg and Sedgwick, 2001; Trowell, 2000; Levinson, 1998: 70–74). Of the Soviet-era statues only the Russian War Memorial remains in its site opposite the US Embassy, where it replaced the fascist *Sacred Flagstaff*. Judith Rugg writes that by decontextualising the sculptures, 'the authorities divested themselves of any obligations to remember and also relieved the viewer of the burden of memory' (Rugg, 2002: 8), echoing Mulvey's idea that a lost history is beyond critique.

Where statues were not relocated, a question arose as to what might move into the vacuum after their destruction. I remember in Bucharest in May 1990, after the fall of Ceauşescu and fighting in the city during which a security presence undifferentiated in the minds of many citizens from that of the old regime was re-experienced, that informal shrines appeared in the street where people had died. But in Romania there were few statues of the kind found elsewhere in the Soviet bloc. Romania saw itself as a non-aligned state with a foreign policy independent of the Soviet Union, and made much of national cultural traditions in open-air architectural reserves, folkloric performances and the painted monasteries of Moldavia. Artists had arrived at coded ways to counter the regime in paintings of church interiors in an atheistic state, or rural villages when these were being demolished. After the end of the regime the codes were superfluous.

The old regime's monuments were buildings rather than statues, especially those of the New Bucharest built after Ceauşescu's visit to Phenian in North Korea.

A debate took place in Bucharest in 2005 as to whether The People's House, Ceauşescu's centrepiece to his new city, should be used as an art museum.

Salecl recalls from a similar debate after 1989: 'Some people insisted that the palace had to be demolished; others proposed that it become a museum of the communist terror; still others suggested that it be transformed into a casino' (Salecl, 1999: 100). As I write the Millennium Dome in London is similarly a potential casino site. For Salecl the new Bucharest represents Ceauşescu's effort to materialize a 'psychotic delirium' (Salecl, 1999: 100) in his later years. The erasure of historical memory in building the new Bucharest on the site of demolished churches, she argues, was in keeping with the grandiose gesture of making a new city. Yet Salecl says that Ceauşescu produced a traumatic disruption 'which now invokes the memory of the lost past' (Salecl, 1999: 102). She observes also that to argue for the retention of statues after the fall of such a regime assumes that 'the current and former rulers do not differ in how they deal with historical memory' (Salecl, 1999: 99). Accepting that erasure may romanticise the past, she argues that at times such erasures are vital: 'If we take the case of post-Hitler Germany, one does not expect to see the Führer's pictures in public places' (Salecl, 1999: 99).

Disgraced monuments

Michalski notes that photographers find evocative images in the decline of monuments:

> Both the process of destruction, displacement and dismantling, which pits human beings and technical appliances against silent, pathetically immobile statues, and the subsequent acts of storage or disposal of broken parts are eminently photogenic and have a surrealistic tinge.
>
> (Michalski, 1998: 148)

Ruins have a power of evocation, as Albert Speer saw when he proposed to Hitler that the Third Reich should construct buildings to ruin well. Only the ordinariness of discarded statues undoes their power. Seeing rows of Lenins in store in Russia as the Soviet Union was fragmenting into its constituent republics, Mulvey and Lewis cite the uncanny:

> Their disgrace and removal may encapsulate, as image and emblem, the triumphal overthrow of an *ancien régime* . . . their ultimate fate raises questions about continuity and discontinuity, memory and forgetting, in history; about how, that is, a culture understands itself across the sharp political break of revolution.
>
> (Mulvey, 1999: 220)

Case study 7.1

From Jeanne Hyvrard, *Les Prunes de Cythère*, translated by Jennifer Waelti-Walters

The mad go in pairs in the alfalfa. Their bells can be heard under the acacias. She fell in the gravel pits one day when she was singing there. Or maybe in the iron stairway that led to the head doctor's office. And yet I had told you to put her in a strait jacket. But me, I want to run, laugh and jump. Saint James's Madness, the mad walk in pairs in the lanes through the wood. Sometimes, one passes someone on horseback. Doubtless, the one who was coming to meet me in the meadow when I was twenty. One day, my horse let go of the bit and you (masc. sing.) dismounted to come to my aid. Your boots were in the grass. The meadow smelled of hay and gentians. Wild blackberries. Raspberries of my only love. Our bodies entwined in the hollow of the rocks. Dandelion. Lady's slipper. Coltsfoot. New flowers of unknown embraces. We've been galloping for days over partly cleared ways. It's the rutting season. We run on the paths to meet the mountain. You are my only love, says the man kissing her in the stable. The smell of horses in the past perfect. The smell of droppings in the future perfect. The poverty of words to smell colours. How many words left to invent to say the cut hay. The imperfect and the pluperfect of the forest, over us like a coat.

But, from hearing you so much I have started to speak patois. Well, said the mother. She is speaking finally. Maybe we will get her to make progress. One doesn't say runned but ran. Poor little thing, she'll never come to anything. She makes so many mistakes. The deformed mouth of the words she can't pronounce. Do you realize my daughter speaks black. Me, a white woman. What will become of me? My little girl. Make an effort. Think of me. Of your father.

(Hyvrard, 1975: 176, cited in Waelti-Walters, 1996: 23)

Comment

Interpreting Hyvrard for anglophone readers, Jennifer Waelti-Walters (1996) argues that the style of her writing embodies a use of chaos theory to relate the systematic economic and cultural denigration of others by the dominant, global economy and its political apparatus: 'The study of chaos has generated ways of coping with vast quantities of information, with the apparently irrational, the unpredictable, jagged edges and sudden leaps' (Waelti-Walters, 1996: 39–40). Hyvrard sees that her novel is an economist's report on Haiti. The narrator's voice is not Hyvrard's but that of a literary character, yet when the book was reviewed in France critics assumed Hyvrard to be Haitian – she is French – and to use a patois which was her own – it isn't – rather than adapting a European voice to conditions in which otherness becomes abjection.

In *La Pensée corps* (1989), Hyvrard proposes a series of redefinitions of the sacred, and questions the idea of a concept. This she replaces with the term 'encept' as a more open and porous way of voicing feelings about the world and human presence in it; while we need ideas to think we are also captured by them, so that the encept simultaneously encompasses both origins and ends, while the concept isolates that with which it deals. Thus, 'Enception is a thought process the thinker is part of as it happens' (Waelti-Walters, 1996: 130). Hyvrard references chaos as well as chaos theory: 'Chaos is the organisation of the world, not its order' (Hyvrard, 1989: 192, in Waelti-Walters, 1996: 130). In premodern Europe chaos denoted an unbounded domain outside the known world. It is not a disordering principle but a space in which formation has not taken place or has dissolved, a grey area of origins and ends in which neither can be distinguished. Chaos is also, I think from a superficial reading of Waelti-Walters's reading of Hyvrard, seen by Hyvrard as a location of authenticity: 'Conception thinks the sacred, Enception is itself of the sacred' (Hyvrard, 1989: 192, in Waelti-Walters, 1996: 130). There is no escape, however, from language as mediation and displacing agent. I find Hyvrard's writing difficult, though the identification of a means of writing which does not bind what is being thought by saying it, but a way of finding what is thought, is attractive – and vital to an understanding of social transformation in as much as revolt which rests on a prefigured concept of a new society will reproduce the old society in which the concept was formed. To break the language alludes to a rupture of historical continuity, like smashing statues. Trinh T. Minh-ha writes, 'Every voyage is the unfolding of a poetic. The departure, the cross-over, the fall, the wandering, the discovery, the return, the transformation' (Minh-ha, 1994: 21). But as Catherine Belsey reminds me (see Chapter 2), the protest is as cultural as that which it protests against.

The ambivalence of the situation mirrors that of public monuments in general, which Lewis writes of as covering up crimes against the public in a history of repression. Lewis continues that, 'when the symbolic order is thrown into crisis, the public monument's semantic charge shifts and the work becomes less heroic in form but rather begins to take on the characteristics of a scar' (Lewis, 1991: 5, cited in Mulvey, 1999: 223). Mulvey notes that the plinth on which a statue of Felix Dzerzhinsky, founder of the KGB, stood before 1991 remained empty after its removal in the Gorbachev period, while in the same square facing the Lubyanka prison a stone from the Solvetsky Gulag was positioned as a monument to the victims of repression: 'Without representing anything in itself, the stone became invested with a collective memory which had not previously found expression in the public space of the city' (Mulvey, 1999: 224). Elsewhere, residual socialism left traces. Stones from the Lenin Monument in Berlin (Nikolai Tomsky, 1970, destroyed 1991) were taken to Friedrichshain cemetery and placed on the grave of Rosa Luxemburg.

Stalin World

I turn now to a specific case of displacement: the relocation in 1999 of seventy-five Soviet-period statues from cities in Lithuania to a 200-hectare forest park at Grūtas, near Druskininkai, an hour's drive from the capital, Vilnius. The park is known as Grūtas Park Soviet Sculpture Museum, or Stalin World. It was founded by the Lithuanian entrepreneur (previously chairperson of a collective farm) Viliumas Malinauskas and received more than a hundred thousand visitors in its first year. The monuments are on a twenty-year loan from the Ministry of Culture following an open tender to find a solution to the problem of what to do with them after removal to storage in 1990. The park opened formally on 1 April 2001 and has a rustic restaurant in which the service is Soviet-style and mushrooms feature in most items on the menu. A field gun is parked in front of it. There is a souvenir shop, a picture gallery, a museum, a small zoo and children's play area. The souvenirs include vodka glasses bearing the faces of Lenin, Stalin and other Soviet heroes, reproduction Soviet badges and red T-shirts. I remain unsure what to think about the statues, except that they are decontextualised in a way that seems to me akin to that of art in white-walled museums of modern art, the walls replaced here by the green-grey forest trees. Soviet music broadcast by loudspeakers evinces manufactured jollyness, and for me an uncanny and fake nostalgia.

That the old regime was not utopian is signified by the relocated watch tower and wire fence. At the entrance to Stalin World is a train with closed cars used for deportations in the Soviet period. The park's literature states, 'One can hardly imagine the extent of horror when in "prisons on wheels" stuffed with people without the most essential sanitary conditions thousands of totally innocent people were deported daily' (Grūtas Park, 2002: np). The memorial aspect is not hidden at Stalin World. Its publicity states:

> This is not a show park. It was created on the basis of a Siberian concept supplemented by nature, wooden paths and fragments of concentration camps. This is a place reflecting the painful past of our nation which brought a lot of pain, torture and loss. One cannot forget or cross out history whatever it is.
>
> (Grūtas Park, 2002: np)

The reality of that history, regarded by many Lithuanians today as a period of foreign occupation, is brought out in the following tourist guide extract:

> The killing Fields:
>
> The remains of prisoners who were shot have since been found buried in Tuskulenai Park, a popular picnic spot in Vilnius until 1994 when the KGB graveyard was discovered. The park is now known as the 'killing fields'.
>
> The remains belong to partisans who put up an armed resistance against Soviet rule from positions in the surrounding forests. The KGB shot them from 1944 to 1947. Of the 708 bodies discovered, only 43 have been identified.
>
> (Williams, Herrmann and Kemp, 2003: 332)

These remains are currently kept in cells at the Museum of Genocide Victims in the site of the previous KGB headquarters at Lukiskiu Aikste, where a large bronze statue of Lenin, now at Grūtas, once stood (fig. 7.3). Another Lenin is shown in the brochure with a wedding couple standing on the plinth. It says, 'The exposition contains fragments of the concentration camps in Siberia'; and 'Children will not be bored at Grūtas Park: Here you will enjoy our playground

Figure 7.3 *Grūtas, statue of Lenin (photo M. Miles)*

Case study 7.2

From Jyoti Hosagrahar, *Indigenous Modernities*

Before the 1857 insurgency British civilians resided within the walls in the northern and eastern parts of the city. Subsequently, they moved out of the walled city to establish a separate European quarter, the Civil Lines to the north. In staying outside the walled city the new rulers lived away from all that the local people perceived as urbane. Yet, the locus of control had moved from the city and its *haveli* to the British quarters outside. Like the city, the *haveli* and their owners had become subservient to the Civil Lines.

The green and airy landscape of the European enclave was in contrast to that of the walled city. Here, British military engineers laid out a settlement of broad regular streets and tidy lots amidst fields and orchards. The bungalows for Europeans were low, sprawling structures set in the midst of gardens on vast lots. The Civil Lines had houses that were set back from the streets and isolated from each other by greenery. The layout was intended to recall English country life and represented to the Europeans the best principles of refined and healthful living in clean and airy surroundings.

In reality, the identities of the *haveli* landscape and that of the bungalow interpenetrated each other, Delhi was not alone in having a European quarter that suggested a suburban model that was the very antithesis of life in the indigenous city. Beneath the apparent opposition, however, was a charged interconnection between the two spaces. The Indians did not passively accept the dualities. The residents of Delhi responded to the new model of urban life in a variety of ways: by disdaining and rejecting, mocking and mimicking, participating and conniving, and learning and accepting. The development of two seemingly antithetical house-forms was deeply inter-related.

(Hosagrahar, 2005: 32–33)

Comment

The space of a city which is the centre of an imperial realm witnessing major insurgency (such as the Indian Mutiny of 1857, to use its colonial name) can become a place of historical forgetting. If Ceauşescu's New Bucharest was an erasure of history to blight popular imagination, the new white district of Delhi after 1857, the Civil Lines, was a removal to open ground of a social group for whom the rupture of historical continuity – a continuity of their making, however, first in the trade of the East India Company and later in military control and political negotiation and subcontracting of power – was traumatic. Hosagrahar notes that the British saw the insurgency as a conspiracy led by the King of Delhi. Just as memory had to be cleansed so the new urban landscape was to be purged of dirt, disease and disloyalty. Hosagrahar remarks, 'The very streets whose mansions and bazaars had earlier enchanted British travellers with their picturesque qualities were now denigrated as unsanitary and a danger to public health' (Hosagrahar, 2005: 86–87). Several districts of the native city were consequently purified by selective demolition. The Civil Lines did not conform to English taste, being like encampments laid out by engineers, yet were built as if in an English suburb with its rustic asymmetries and excess of space in non-productive gardens. This projects an orderliness taken to signify a future guarantee of security. The crowded streets of the native city, in contrast, and the hidden spaces of the courtyards of its houses and *haveli*, received a negative projection in which they corresponded to the inner-city, low-class areas of London or Manchester which Victorian reformers sought to improve – to lessen the threat of revolt which remained a spectre in official imagination throughout Victoria's reign. Today a new cleansing operates as gated apartment buildings come to Delhi (Seabrook, 1996: 210); and a re-emerging, commercial Indian upper class is offered the products of an image of security in the guise of English houses in the country. On the outskirts of Bangalore, Anthony King reports, the development of Oxford Impero offers a distillation of 350 years' worth of architecture in 550 premium apartments (King, 2004: 133–134) on a greenfield site. But cultural currents operate in all directions, and the transplanting of English suburbia to Delhi is different only in detail and context from the use of stone cladding on British inner-city housing terraces as a mark of ownership – one of few visible recodings available in these circumstances, borrowed from the nineteenth-century use of stone in bank buildings. That this is derided by the cultured class demonstrates the role of visible signs in a continuous fragmentation of urban environments along lines of class or other division. The ordinary detail of the built environment, then, is itself a monument by other means and, like public monuments of more conventional kinds, subject to recoding.

with amusements and a mini zoo' (Grūtas Park, 2002: np). I was conscious that academics are also tourists while visiting Stalin World; and that the brochure is in English for tourist consumption. Yet most visitors are Lithuanians, some of whose comments are translated: 'I express true gratitude'; and, 'The future generation who will study history in your park will be grateful' (Grūtas Park, 2002: np). A selection of letters for and against is translated, too: 'Blacking other people is not the best way to stay "white" . . . It seems that money is your only value. And it's a shame because your start was very promising' (Grūtas Park,

2002: np). Another argues that retrospect depends on access to history, affirming Mulvey's position that the relics of a regime allow a coming to terms with its history.

The park's memorial aspect co-exists with its function as a place where families enjoy days out, and a tourist attraction. It is contextualised by an antagonistic attitude to the Soviet liberators who did not leave after 1945; and by the election in 1992 of the former Communist Lithuanian Democratic Labour Party and sale of Soviet-era memorabilia in a street market in Vilnius. Stalin World is mentioned in Michael Connelly's thriller *Void Moon* (2000: 407), and seems to be at the intersection of cross-currents of post-socialism, Europeanism and resurgent nationalism. Such currents can be difficult to negotiate. Susan Buck-Morss remarks that, in a Soviet film symposium in Moscow in 1991, a young director's film *Stalin in Africa* (a spoof about a Stalin double) was met with embarrassed silence, while Moldovan delegates called for a national cinema even though their politics were at odds with the realities of global film production and distribution (Buck-Morss, 2000b: 245). The Moscow symposium was over-shadowed by a KGB attack on the radio station in Vilnius on 13 January 1991 in which 580 people were wounded and 14 killed. A large wooden cross is sited on the hillside leading up to the radio station now. Buck-Morss records that the first Gulf War began three days later, demonstrating that, 'on both sides, the machinery of violent state power could be reassembled despite the absence of Cold War justifications' (Buck-Morss, 2000b: 247).

Vilnius today

The new government in Vilnius has commissioned a statue of Lithuania's first king, of whom there is no visual record. In 2005, the medieval palace next to the cathedral was being rebuilt using archive material to reconfigure its design (despite that at no historical moment did the palace and the cathedral stand next to each other, one destroyed when the other was built). Support for the con-struction of monuments of national identity outstrips that for contemporary cultural production. Art exhibitions flourished initially after 1990 through external, international support. Vilnius has shops representing global fashion chains, but a bridge across the Neris river, known as the Green Bridge, which connects the centre to the north bank's Soviet-period buildings and a shopping mall and international chain hotel, retains its socialist realist sculptures – two figures stand at each corner: workers, peasants, soldiers and airmen. In 1995 Lithuanian artist Gedimas Urbonas placed mirror-glass cubes over the heads of each figure on the Green Bridge, provoking a heated debate as to the future of the only Soviet monument still occupying prominent public sites. Laima Kreivyte (who was present outside the radio station in 1991) describes the project as 'one of the best examples of how public art . . . has moved from the representational to the discursive' (Kreivyte, 2005: 129).

Conclusion

What is it appropriate to do with the monuments of past regimes? Should they be junked, given to historians to keep, left in place as reminders of a history which cannot be denied, or recoded? As an example of the latter I cite the *Mother Russia* monument in Kaliningrad. Here, in a Russian enclave between Lithuania and Poland with a German and Hanseatic cultural history extant in the remains of towers, gateways and a network of underground passages and bunkers holding stolen museum treasures (Sezneva, 2002: 58), the official version of the statue's meaning contrasts with its popular appropriation. Olga Sezneva relates that while it is described in a Soviet-period guide to the city as representing in a downward pointing finger that the land belongs to Russia, 'people's imaginations reworked the memorial in another, significantly different way . . . accidentally or not, a finger of the right hand, directed down at a certain angle, appears as a particular attribute of masculine anatomy' (Sezneva, 2002: 59). The statue replaced a Stalin demolished in 1957, and the figure had originally been designed as a representation of the Greek fertility goddess Demeter holding a pot of honey. Sezneva continues that 'collective imagination together with irony and mockery led to practices that subverted the Soviet state's apparent representational hegemony' (Sezneva, 2002: 59). This included popular memory of the German cathedral demolished in the Soviet era.

I never lived under the Soviet system, and sentimental regret at its passing is out of place. But, as Nancy Fraser argues, there was an opportunity in the 1990s to reformulate socialism in a democratization of culture and everyday life moving towards radical democracy. She continues,

> What happened, alas was nothing of the sort. Far from helping to expand the socialist vision, postindustrial movement politics effectively displaced it. For me . . . the greatest disappointment . . . was the fading away of an entire grammar of social conflict. The socialist worldview, centred on the reorganization of labor and the redistribution of wealth, ceased to supply the terms of political contestation – not only in East-Central Europe but throughout the world.
>
> (Fraser, 2001: 201)

The disgraced monuments indicate little of that grammar. To erase them would remove a sign that it existed. To do so would also re-enact the shift of power.

Case study 7.3

Figure 7.4 Arcosanti, Arizona (photo M. Miles)

Comment

I went to Arcosanti in 2004, researching utopian settlements for a forthcoming book. It was a long journey, using a lot of my notional carbon allowance. This allowed a disconnection from my usual environments, and from the relatively European environment of Cambridge, Massachusetts, where I spent a few days looking at a co-housing project before flying to Phoenix. Arriving via an airport pick-up service at a gas station in the desert, surrounded by a landscape like those I remember from cow-boy films viewed in my childhood, I was met by one of Arcosanti's residents as arranged. On arrival I was led to a simple, white en-suite guest room with a view over the desert. I spent four days at Arcosanti, walking a lot, swimming in the almost unused open-air pool, eating in the communal restaurant serving the settlement's own organic produce, trying to understand what Arcosanti was. Materially, it was concrete, in rectilinear modernist style and vast domes shading public spaces. It was also green, with cypress trees (watered by sprinklers) lending the hilltop site a Tuscan association – though in Tuscany cypresses tend to denote a cemetery or a local Calvary. Conceptually, it was architect Paolo Soleri's realisation in progress of an experimental city for five thousand people, isolated in the desert and at the time housing and for the most part employing a community of seventy-two. Soleri himself – who once worked for Frank Lloyd Wright but, I was told, had been sacked for being late back from vacation – lives on a nearby ranch. The construction continues slowly, assisted by students who do summer courses which include hands-on construction (or labour for which they pay the proprietor in a reversal of capitalism's usual arrangement) as well as slide talks on Soleri's theory of Arcology. I asked one resident why Arcosanti did not generate its own energy instead of taking it from the grid, given the perpetual sun and frequently strong winds, but his answer was cautious, suggesting this was not on Soleri's agenda but might be introduced at a later date – by implication after his death. The settlement hosts cultural events such as jazz festivals, which are well sup-ported, and makes an income of around a million dollars annually from selling bronze and ceramic bells. A 'legendary' source on Soleri is said to establish him as 'one of the most innovative minds of our time' (www.arcosanti.org/media/city.htm – citing Soleri, 1969). I found Arcosanti uncanny, a pseudo-Tuscan hill town in Arizona, a monument to its founder but equally to a utopianism which, in international modernism, relied on preconceptions and was thus itself part of hegemony that modernist avant-gardes sought to replace.

The main points I have made are that there is a need in certain conditions to re-enact a shift of power; that this may be articulated as a destruction of iconic representations of a past regime, or less dramatically expressed in unofficial memorials such as the stone opposite the Lubyanka prison in Moscow; that to leave the monuments in place, or re-place them, may enable a coming to terms with a history's negative and positive aspects; and that the official rhetoric is always undermined informally. But if liberation is a real possibility there are other questions than what to do with statues. Student activists in 1968 engaged in new music, free love and alternative social arrangements. Julia Kristeva, arriving in Paris from Bulgaria in 1965, recalls that '[1968] was a worldwide movement that contributed to an unprecedented reordering of private life' (Kristeva, 2002: 18). The revolutionaries of 1968, amid cross-currents of sexual liberation, multiculturalism and Maoism, did not make statues but situations. When the Situationists painted Henri Lefebvre's words 'Under the paving stones, the beach' (*sous les pavés, la plage*) (Pinder, 2005: 238, fig. 8.1) on the streets of Paris they meant a real utopia was concealed within the routines of everyday life. Perhaps liberation is in intimate encounters:

> The will to join private jouissance to public happiness isn't necessarily communitarian. It responds to the need for social ties, which . . . is a condition of psychic life. The resurgence of religions and sects reflects this . . . With the difference that in the French movement, sexual experience was thought through, analyzed and theorized by way of psychoanalysis and Enlightenment materialism. This approach favoured . . . a kind of radical atheism which is perhaps the only non-vulgar or dogmatic form of atheism: exhausting transcendence in its own terms by liberating speech through eroticism.
>
> (Kristeva, 2002: 40)

8 Cultural identities

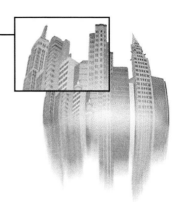

Learning objectives

- To gain insight into the uses of culture in identity formation
- To differentiate between cultural production and the cultures of daily life, while seeing how they sometimes fuse
- To understand how cultural work operates in non-affluent urban conditions

Introduction

In Chapter 7 I sketched a context of globalisation, and investigated some of the intersections of cultural frameworks in post-Soviet states today. In this chapter I turn to the formation of identities through cultural work, by which I mean mainly work in the arts but outside art's conventional sites, similarly outside the sites of consumerism and bordering on everyday life. I focus on *Capital do nada*, a multi artform project in Marvila, a multi-ethnic social housing district of Lisbon, in 2001. I contextualise this by reference to the sociology of consumption and emphasis on consumption as a means to identity formation (extending the discussion of consumption in Chapters 4 and 5). I doubt the hypothesis that lifestyle consumption is a means to identity formation, and look at Marvila as a zone offering little access to such consumption. I describe Marvila, outline the project *Capital do nada*, and examine two projects within it: *Belcanto* by Catarina Campino, a performance project in five locations; and *porque é existe o ser em vez do nada?* by José Maças de Carvalho, a photographic project using mass media sites and imaging techniques. Finally I reconsider issues of power and empowerment raised by the projects, citing Hal Foster (1996) and Irit Rogoff (2000). In case studies I cite an extract from Arundhati Roy's *Power Politics*

(2001); and a text by critic bel hooks which addresses the idea of an aesthetics of everyday life. My final case study uses an image of puppet making at the Social Work and Research Centre (SWRC), or Barefoot College, in Tilonia, Rajasthan, India.

Contexts

Identity and consumption

During the 1990s various areas within the social sciences began to adopt new perspectives. One of these followed the translation into English of Henri Lefebvre's (1991) theories of space (see Chapter 1) – though Rob Shields argues that geographers overlooked Lefebvre's theory of moments of liberation within the routines of everyday life (Shields, 1999: 58), an earlier formulation out of which the spatial theory grew. The impact of feminism brought questions as to the privileging of visuality and subordination of the first-person voice (Massey, 1994), echoing a focus on the right to a voice in post-colonial studies (Minh-ha, 1994; Spivak, 1988). This is echoed, too, by planner Leonie Sandercock (1998) in a chapter of *Towards Cosmopolis* titled 'Voices From the Borderlands: The Theory That Difference Makes', in which she states, 'If we want to achieve social justice and respect for cultural diversity . . . we need to theorize a productive politics of difference. And if we want to foster a more democratic, inclusionary process for planning, then we need to start listening to the voices of difference' (Sandercock, 1998: 109). These perspectives – if that is an appropriate term for work which decentres its field – brought the non-affluent world, and zones of non-affluence within the affluent world, into academic study, but not as early ethnography had done as a generalised Other constructed to show Western supremacy. In this general context, interest in consumption as a means to identity formation (Shields, 1992; Nava, 1992; Lury, 1996; Miles S., 1998) emphasised the empowerment of consumers, particularly women (Nava and O'Shea, 1996; Ryan, 1994; Friedberg, 1993) and new urban publics using consumption or cultural spaces in their own ways (Miles, 2003; O'Connor and Wynne, 1996).

Martyn Lee argues that cultural studies reads cultural forms as sites in which social meanings are contested, and continues:

> When applied to the topic of consumption, this observation suggests that consumer goods lead a double life: as both agents of social control and as the objects used by ordinary people in constructing their own culture. Ideologically and aesthetically contextualised by advertising, design, marketing and other promotional forms, commodities are to be considered as 'texts' inviting certain preferred forms of reading and decoding which aim to reproduce dominant

social relations. But in their role as cultural artifacts, consumer goods become important material and symbolic resources with which ordinary people reproduce their life and their patterns of life.

(Lee, 1993: 48–49)

This assumes consumers are able to read commodities discerningly.

Minjoo Oh and Jorge Arditi (2000: 71) see a shift of focus from production to consumption in post-industrial societies, and for most writers consumption is key to the definition of contemporary urban living even when its alignment to the aesthetic values of culture is questioned (Warde, 2002; Miller, 1998). To me it seems that a cultural critique of consumption brings to it the autonomy associated with modern art. This is problematic – the autonomy is illusory in both cases – but admits an element of critical distancing within the notion of autonomy and potentially open to disaggregation from it. Lee differentiates the concepts of a market-led economy and cultural mediation of consumption: 'the culture of consumption is never . . . a mere symbolic echo or the purely functional realisation of product positioning by advertising and marketing strategies' (Lee, 1993: 49). A cultural critique of consumption, then, moves to a model of dialectics as the effective means of understanding consumption (Miller, 2001). Lee clarifies this when he writes that 'consumers have clear limits placed upon the range of meanings and uses which they may assign to commodities', and that the design and symbolic contextualisation of commodities is likewise 'structured by the lived meanings and uses of commodities as they have passed over into the status of cultural objects in everyday life' (Lee, 1993: 49). Theodor Adorno's critical insights into mass culture (Chapter 5) are as relevant to consumption in a wider field as to that of broadcast media on which he focused.

A poster for graduate unionisation in New York University states, framing an image of a woman taking a knife and fork to a volume of Marx, 'I can't eat cultural capital' (in la Berge, 2006: 63). On the other hand, presumably the union reflects at least distantly the idea of class organisation which is one element of Marxism. Matthew Daunton and Martin Hilton, in *The Politics of Consumption*, read all consumption as politicised while the cultural politics of style which they attribute to cultural studies shows 'collective or subcultural identities emerging out of class- and youth-based protest against the state' (Daunton and Hilton, 2001: 11). This leads to the question as to whether there are means to articulate demands for identity or a voice in the determination of a society's future shape and values outside consumption.

Case study 8.1

From Arundhati Roy, *Power Politics*: 'The Reincarnation of Rumpelstiltskin'

Remember him? The gnome who could turn straw into gold? Well, he's back now, but you wouldn't recognize him. To begin with, he's not an individual gnome anymore. I'm not sure how best to describe him. Let's just say he's metamorphosed into an accretion, a cabal, an assemblage, a malevolent, incorporeal, transnational multi-gnome. Rumpelstiltskin is a notion (gnotion), a piece of deviant, insidious, white logic that will eventually self-annihilate. But for now, he's more than okay. He's cock of the walk. King of All That Really Counts (Cash). He's decimated the competition, killed all the other kings, the other kinds of kings. He's persuaded us that he's all we have left. Our only salvation.

What kind of potentate is Rumpelstiltskin? Powerful, pitiless, and armed to the teeth. He's a kind of king the world has never known before. His realm is raw capital, his conquests emerging markets, his prayers profits, his borders limitless, his weapons nuclear. To even try and imagine him, to hold the whole of him in your field of vision, is to situate yourself at the very edge of sanity, to offer yourself up to ridicule. King Rumpel reveals only part of himself at a time. He has a bank account heart. He has television eyes and a newspaper nose in which you see only what he wants you to see and read only what he wants you to read . . .

Listen carefully. This is most of the rest of his story. (It hasn't ended yet, but it will, it must.) It ranges across seas and continents, sometimes majestic and universal, sometimes confining and local.

(Roy, 2001: 35–36)

Comment

Roy is writing in protest against the construction of a series of dams in the Narmada Valley, of which around thirty are planned, in construction or already built (in 2006). Having gained international attention with her Booker-prize-winning novel *The God of Small Things*, Roy uses her status to draw attention to the destruction of natural and social environments by schemes in which Western companies have major interests. She notes that India repays more in interest and capital to the World Bank than it receives, having to incur new debts to pay old ones. She observes also that export credit agencies enable transnational companies to be certain of payment for large capital schemes such as the dams. Roy (2001: 62–74) gives a set of facts: the dam industry is worth $32 to $46 billion a year. It needs new markets because dams are out of favour in the affluent world. India has the third largest number of dams in the world. Forty per cent of the big dams in the world are in India, where they are presented by politicians as national symbols. Ninety per cent of big dams in India are for irrigation, seen as food security, but produce less than 10 per cent of Indian grain; 350 million Indian citizens live in poverty. Around 56 million people may be displaced by big dam schemes, many of them Dalit and Adavasi (the poorest of the poor). The Bargi Dam completed in 1990 cost ten times its budget and when its sluice gates were opened (at night), 114,000 people from sixty-two villages were displaced and twenty-six resettlement colonies submerged. The Sardar Sarovar Dam – financed by a $450 million loan from the World Bank – has become 'the showcase of India's Violation of Human Rights Initiative' (Roy, 2001: 72). India shows a portrait of M. K. Gandhi on its banknotes, a key point in his political programme being that India should not industrialise on a Western model but preserve its village-based traditions of farming and manufacture. That does not mean it should have no industry or machines, but that industry should be used selectively. As Glyn Richards summarises, 'Machines that lighten the burden of millions . . . are welcome, but labour-saving machines that put thousands out of work are rejected' (Richards, 1991: 122). The dams of which Roy writes are mega-machines. She recounts how her political writing has brought objections, that as a writer she is required to 'be ambiguous about everything' (Roy, 2001: 12). Her right to free speech is guarded under civil law yet she was prosecuted in 2001 for committing contempt of court by organising a demonstration in protest at the Sardar Sarovar Dam (Roy, 2001: 87–103). The Introduction to volume 4 of *Sarai Reader* (see case study 9.2) states:

> This book carries with it the unreasonable expectation that the myriad realities that seek to make themselves known in a world that usually silences them, will find voice, and that we will all find the patience to listen very carefully. These are times for sober reflection.
>
> (*Sarai Reader*, 4, 2004, www.sarai.net/journal/reader4.html)

Sober reflection is the prerogative of writers and academics but denied people whose homes are flooded in the night to meet the requirements of global capital. Yet the expectation is logical, in a cold technocratic way. The erasure of myriad voices is the means by which globalisation controls the world. It may seem like tilting at windmills to oppose it but the walls of Jericho are said to have fallen when trumpets were sounded for long enough in procession round them. Part of the responsibility of cultural producers, for Roy, is to show conditions as they are, so that rational responses can occur.

Participatory arts

One history of modernism, advanced by the critic Clement Greenberg in the 1960s, states a trajectory the apex of which is pure abstraction. Another, found in the writing of Raymond Williams (1989), states confused or contradictory political positions aligned with a growing autonomy in the means of art's production and distribution. Greenberg's trajectory shares the progressive and reactionary dimensions alike of early ethnography's trajectory of universal human development (see Chapter 2): Greenberg is progressive as a leftist critic identifying a succession of movements towards a state of abstraction he equates with freedom; and reactionary in that the trajectory universalises the condition of elite culture in the mid-twentieth century as the ultimate point of art's development, and affirms his own position as arbiter of such judgements. The contradictions drawn out by Williams in Futurism remain interesting, however, because they suggest that currents of autonomy, contingency and utopianism mix in art. I would see this as most likely in art which seeks to operate outside market values or to engage in social issues. As in development work in non-affluent settings, the question is how far the process is handed over to non-artists (audiences) or the artist's status as expert professional retained and the project added to the artist's portfolio.

The issue is only partially open to resolution, as democratic societies aspire to a negotiated position between freedom and unfreedom (Laclau, 1996: 19); but to me it is crucial because, unless power is handed over, the power relations governing existing conditions are reproduced, and the only way in which such reproduction can be avoided is when the means embody rather than signal the desired ends. Michel Foucault concludes:

> Domination is in fact a general structure of power whose ramifications and consequences can sometimes be found descending to the most incalcitrant fibres of society. But at the same time it is a strategic situation more or less taken for granted and consolidated by means of a long-term confrontation between adversaries.
>
> (Foucault, 2000: 26)

Which suggests that means determine ends and that oppositionality is not the only possibility of struggle.

In the 1990s artists whose work in non-gallery sites represented a desire to engage with social and environmental questions and work collaboratively with non-art publics articulated a counter-argument to that of public art in urban redevelopment. The critical debate surrounding this departure focused on an art of public issues (Raven, 1993), participatory processes (Felshin, 1995; Lacy, 1995) and locality (Lippard, 1997). Although public art had been a departure from the gallery and refusal of art's commodity status in the 1970s, by the 1990s its allegiance to market interests and the expediency of cultural policy (Yúdice, 2003) suggested a return to mainstream structures. Whether public artists understood public space as a site of multivalent contestation, or saw it simply as a site in which to put their work, was also questioned by some critics (Deutsche, 1991; Phillips, 1988). At the same time, outside public art dissident groups developed practices characterised by engagement in politicised situations.

As an example I cite the project *Intervention in a School* (1995) by Vienna-based collective WochenKlausur, which consisted in the participatory re-design of a school classroom with clusters of semicircular seating to replace rows of desks facing the front. Their intervention in the design of space was seen as facilitating new attitudes to the production (rather than reception) of knowledge. In a range of projects WochenKlausur use the critical distance of art to foster approaches which reconfigure social structures through dialogue. As Grant Kester writes in *Conversation Pieces*, there can be difficulties with 'advocacy projects that benefit disempowered populations' when outsiders speak for them, but 'Wochen Klausur appears to be cognizant of this problem' (Kester, 2004: 100–101). Kester cites a member of the collective, Wolfgang Zinggl: 'The context of art offers advantages when action involves circumventing social and bureaucratic hierarchies . . . to accomplish concrete measures' (in Kester, 2004: 101). In *Intervention to Aid Drug-Addicted Women* (Zurich, 1994–95), WochenKlausur invited sex workers, journalists, local politicians and activists to join a series of dialogues on a boat on Lake Zurich. The space of the boat was metaphorically a space of autonomy but also of critical distance allowing participants to adopt other than predictable attitudes, and reach 'a modest but concrete response to the problem' (Kester, 2004: 2) in the form of a hostel for sex workers, since completed. A range of other groups could be cited, with varied emphases on social and environmental justice (Kester, 2005; Strelow, 1999), provoking civic contestation (Miles, 2004: 147–179) or feeding dissent in cultural circles (Miles, 2005). *Capital do nada* took place in this context, in the year in which Porto – Portugal's second city – was a European Capital of Culture (see Chapter 6). Its disdain for awards has a Dadaesque feeling, in contrast to the conventional model of bringing culture to a mass public. Rather than cater for preconceived needs the programme intervened to provoke responses leading to dialogues that had a potential to open unexpected inquiries and produce unforeseen insights into the social life of a neighbourhood.

Case study 8.2

From bel hooks, 'Beauty Laid Bare; Aesthetics in the Ordinary'

At the outset of the contemporary feminist movement there was significant interrogation of consumerism, of women's addiction to materialism, and of the issue of money, both its distribution along gendered lines and its use . . . An anonymous 'red-stocking sister' made the useful point that feminist discussions of female consumerism would be useful if they began from a standpoint that depicted Americans as mere dupes of patriarchal advertising culture . . . As was the case in black liberation struggles, there was no discussion of aesthetics, of the place of beauty in everyday life . . . Progressive feminist thinkers are more likely to critique the dangers of excessive materialism without discussing in a concrete way how we can balance a desire for beauty or luxury within an anticapitalist, antisexist agenda.

. . .

Most radical or revolutionary feminists continue to believe that living simply, the equitable distribution of resources, and communalism are necessary to the progressive struggle to end sexism while ending class exploitation . . .

Rather than surrendering our passion for the beautiful, for luxury, we need to envision ways those passions can be fulfilled that do not reinforce the structures of domination we seek to change . . . feminist thinkers will begin to engage in more discussion and theorizing about the place of beauty in revolutionary struggle. Many of us who have a degree of material privilege find that sharing resources, sharing objects we find beautiful that enhance our lives, is one way to resist falling into a privatized, hedonistic consumerism that is self-serving. Those of us who engage in barter, conscious gift giving, tithing, sharing of living space and money, celebrate the luxurious if that which we deem luxurious is not acquired by harming others.

(hooks, 1995: 122–123)

Comment

Necessity can be defined as meeting survival needs. Black Asian architect Shaheen Haque writes of her student experience: 'The lectures particularly reinforced the dominance of western art and architecture and portrayed them as the most important influence on architectural design. Other significant forms such as Islamic and African architecture were relegated and merely seen as precursors for the more important western traditions' (Haque, 1988: 56). Things have changed since that was written, but not much. Recognition is a necessity in this situation, and indicates that not all necessities are material. Luxury is exotic amid oppression which denies a person's right to possession of a culture, like something from the *belle époque* or literary fiction. While hooks uses the term to suggest the personal or intimately enjoyed as balance to the public and overtly ideological, it does not cease to be political. The personal is always political just as the political always has had personal impacts. One reading of the text, then, is that barriers between personal and political categories, as between high and everyday cultural production and reception, or between culture as the arts and culture as a way of life imbued with its own aesthetics, are collapsing. Another reading might be that the cultural has been largely ignored in radical thought, and in feminism in particular. Culture was often viewed with suspicion in debates on class struggle, but not by Raymond Williams, Stuart Hall or Richard Hoggart in the formation of cultural studies. In cultural studies, culture was a ground of contestation and antagonism, as it is for hooks. It is in language, notably, that culture and cultures are brought into focus, put in the spotlight and shown to perform for the dominant force in a society. For example, should I wish to say I have equal ability to use my right and left hands I say I am ambidextrous, which means not exactly that I have two right hands but that both hands are classified as right hands (dextrous not sinister). As Jennifer Waelti-Walters writes, hooks 'sees language as a site of struggle' (Waelti-Walters, 1996: 153 – see case study 7.1), while Adrienne Rich writes that 'This is the oppressor's language, yet I need it to talk to you' (cited in Waelti-Walters, 1996: 153). The word *luxury* is no less open to renegotiation in meaning than any other, like class, or power, or capital of the kind you can't eat. Apart from anything else, the passage quoted above is useful in getting away from the ascetic view of alternative lives, and confirms that in everyday practices values other than those of consumerism are expressed. This renegotiation is integral to that of language, which Luce Irigaray develops as an art form (1996; 1999); more prosaically she writes: 'language is not neutral . . . its rules weigh heavily on the constitution of a female identity and on women's relationships with one another' (Irigaray, 1994: 27). The same can be said for visual as for verbal language, for the dominant position of the perspective view and the false coherence of a city's skyline. Sunlight disregards this as it filters through the washing on to the grass of the suburban lawn, or fuels the photogenesis of a plant on the concrete balcony of a social housing block. But sunlight is inanimate. We are not.

Capital do nada

Marvila

Marvila is a blank on the map of Lisbon (literally so in most tourist guides), between the city's nineteenth-century extension and the Expo-98 site. In the twelfth century it was the site of several mosques, in later years of farms, vineyards and olive groves owned by Catholic religious orders. A gunpowder factory operated there in the nineteenth century. Traces of farms remain in old villas and patches of bright green cultivation. Marvila epitomises the periphery to which the poor and migrant are generally displaced, a peri-urban zone of concrete blocks and wasteland visible from the airport road which also contains an area of residual nineteenth-century workers' houses, factories and wash-houses and some self-built housing. The blocks were built in clusters between the 1960s and the 1990s, are zoned alphabetically and house immigrants from Portugal's rural areas and its former colonies of Angola, Moçambique, San Tomé and Cabo Verde. Marvila typifies the transitional landscapes which surround large cities in a lack of distinctiveness, though the blocks were painted in bright colours in preparation for Expo-98 because they could be seen by visitors using the airport road to the Expo site. But Marvila's architectural impoverishment is balanced by the strength of its invisible social architectures – the structures of social cohesion and organisation such as tenants' councils and ethnic associations, and structures of representation mainly through the Portuguese Communist Party. To the south is a zone of redundant docks on the Tejo (replaced by a container port) where several buildings have new cultural or leisure uses, and architects' studios have appeared in an incipient culturally led redevelopment – but this is not Barcelona.

Marvila is a product of planning, unlike the unplanned *favellas* inhabited by migrant groups on the edges of Lisbon. It is multi-ethnic but not cosmopolitan in that there are no ethnic restaurants or designer bars selling imported beers. The nearby metro station is accessed through a shopping mall which opens at 10 o'clock and cannot therefore be used by workers from Marvila going to work in the early morning. In Marvila there are a few restaurants and bars, a small number of shops, several schools and a technical college. Some of the bus stops have shelters but the services are less frequent than those of more affluent districts. The blocks are grouped in micro-neighbourhoods surrounded by quite large tracts of waste ground crossed by highways. The publication which followed *Capital do nada* sums up this landscape,

> Marvila carries the burden of a peripheral place although it is inside the city.
> It is one of those places that survives maybe because of the lightness of so

much sky and of so many open sights that the continuous fabric in the continuous city no longer allows. It lives in an imponderable vastness para-doxically provided by that same burden.

(Extra]muros[, 2002: annex, p. 23)

It is Lisbon's equivalent of the orbital urban landscape described by Iain Sinclair (2003 – see case study 1.1); and of the peripheral zones of Paris of which Stephen Barber writes in *Extreme Europe* (2001).

From a peri-urban viewpoint, framing the views of luxury apartment towers and the district of *La Défense*, Barber sees 'a vast, abandoned expanse of broken glass and graffiti, between two suburbs which barely possess names. All around, motorway overpasses traverse the edges of the city in a fierce, relentless medium of noise' (Barber, 2001: 98). In Vitry – where inner-city communists moved in the 1960s followed by immigrants in the 1970s – he hears 'a vibrant cacophony' of languages from the windows of 'graffiti-screened buildings', observing,

> Every morning and evening, the inhabitants of the blocks face a long walk to and from the distant suburban railway station, to take packed, dirt-encrusted trains into the city, pressed eye to eye, to undertake sweatshop or street-gutter work.
>
> (Barber, 2001: 114)

Marvila has less graffiti but its marginality is constructed in a similar way by geographical isolation, design and lack of amenities. The earlier blocks are of a particularly poor standard, though no worse than project housing of the same period in, say, Baltimore (Harvey, 2000). They functionalise space in a way that denies the capacity for self-organisation of life in the traditional inner-city street with its multiple and overlapping uses (Robbins, 1996). These peripheries are not edge-cities (Garreau, 1991), or non-places – 'spaces formed in relation to certain ends (transport, transit, commerce, leisure), and the relations that indi-viduals have with these spaces' (Augé, 1995: 94) – but sites which identify their occupants as marginal within a European capital city once the hub of an empire stretching from Brazil to Macau in China. But local political action in the early 1990s replaced the alphabetically denoted zones by names – an act of empowerment preceding the intervention of arts workers.

The project

Capital do nada was conceived as a series of cultural interventions to enable dwellers in Marvila to make invisible architectures visible. This is not to say that it sought to encourage belief in them, since those present were active in them and well aware of their benefits. The difficulty is instead that dwellers in marginal zones are denied the resources of a city, and have little or no voice in the determination of its image. Hannah Arendt (1958) saw this kind of invisibility as a painful condition, depriving those disaffected by it of a public realm in which to form self-identities through the perceptions of others. This denial is common to peripheral situations.

Capital do nada was envisaged as a catalyst to self-realisation on the part of the multi-ethnic, working-class and ex-agrarian publics of Marvila. It was initiated by curator Mario Caeiro, artist Daniela Brasil (see case study 6.3), architect Luis Seixas and geographer Teresa Alves. The project team acknowledged previous efforts, as in naming the districts, and worked through extant political structures and social networks. Several representatives from tenants' councils became involved. The President of the municipality, António Augusto Pereira, wrote,

> I witnessed the very beginning of this project . . . and although there have been many obstacles and difficulties, I'm glad to say that the results have surpassed our expectations . . . every day we could read about Marvila in the newspapers and hear about Marvila on the radio . . . Above all, the people of Marvila had the opportunity to stop, look around and comment on their own neighbour-hoods, associations, culture, architecture, urbanism, parks, environment and many other subjects.
>
> (in Extra]muros[, 2002: 4)

The programme began on 1 October 2001 when four thousand people joined hands in a human chain linking the district of redundant factories and small houses to the new zones of Marvila's blocks.

Many of those making this chain were primary- and secondary-school students wearing the project's red T-shirts. Francisco Sousa wrote:

> For me the human chain was a way to show the world that I exist and to share my civic interior with others. This project 'giving hands for nothing' meant to show that we all have the duty as well as the right to citizenship.
>
> (in Extra]muros[, 2002: 123)

And Ana Ruis Dias:

> By participating in the human chain I felt I was having a citizenship act, that I was being a good citizen. I felt happy with myself. I enjoyed it a lot and I feel honoured that I participated in it.
>
> (in Extra]muros[, 2002: 4)

On a similar note, Guilherme Pereira wrote:

> When we gave hands, I felt that we were all united and that we belonged to Marvila. It was fun, there was a lot of people and joy. Everybody participated in this event, from the 6th to the 12th form, the Director, the co-ordinators and teachers. We all 'gave hands for nothing'.
>
> (in Extra]muros[, 2002: 122)

The remarks are selective, included in the programme publication. I believe them to be genuine. From more recent visits I think (but as an outsider, of course) that the programme helped increase the confidence of local and political organisations to put Marvila on the map of Lisbon. A recent publicity leaflet to attract investment and tourism to Marvila states, 'Marvila is a dynamic district' (Marvila, nd). I move now to two projects within the overall programme.

Belcanto

Belcanto was a performance work by Caterina Campino performed in five locations in Marvila: a road bridge over a railway line in Pátio do Marialva (4 October at 8.00 am); a road adjacent to a construction site and municipal building in Marquês de Abrantes, previously Zone Z (9 October at 11.00 am); inside the restaurant Churrascaria Pi-Pi, Marvila (14 October, a Sunday, at 3.30 pm); a playground in a pedestrian precinct in the centre of Marvila (18 October at 4.00 pm); and Parque da Belavista, a large green area on the northern edge of Marvila (27 October at 3.00 pm). At each location a large, white Mercedes van arrived out of which workmen unloaded a concert grand piano which was professionally tuned. A pianist and singer in concert dress arrived next in a black limousine, also a Mercedes, to give seemingly impromptu performances from a mainstream classical and operatic repertoire – Mozart, Tchaikovsky, Rachmaninov, Donizetti, Verdi, but also songs by Gershwin. The performances were co-ordinated by cell-phone and recorded on video. The project diary records, for the second performance, 'These places make up a large part of the . . . urbanisation of the area. They are places full of mud and rubble and where there is constant noise and bustle. The target audience is the "city builders", the Eastern European, African, and Portuguese workers (Extra]muros[, 2002: 180). And for the fourth, 'The children, arriving home from school, came to the playground to play and saw the opera recital. The locals who were at home came out onto their balconies to hear Rodrigo's aria to Don Carlos' (Extra]muros[, 2002: 184). In England this might be an Arts Council project to widen access to high culture by taking it to unconventional places, like giving bibles to animists, affirming the differential status of producers and recipients; or a community-arts project from the 1970s. But the concerts fit more in a Dadaist tradition specific to a European cultural history: cultural shock tactics trading on the bizarre, like the first Dada performances in Café Voltaire in Zurich in 1916. The concerts were not announced and there were no tickets or security guards. They ended as abruptly as they began, though participants in some cases played the piano afterwards. At the railway bridge: 'The sound and the presence of the trains that arrive and depart. The sky and the river in view. Signs saying "Danger/High Voltage!". Wake to the sound of Rachmaninov' (Extra]muros[, 2002: 178). This sounds like Futurism in the 1910s transposed to a deindustrialised society whose members live in a forty-year construction site.

porque é existe o ser em vez do nada?

Turning to *porque é existe o ser em vez do nada?* (*why is there being instead of nothing?*), the project was co-ordinated by artist José Maças de Carvalho with members of tenants' and other associations in Chelas, Armador, Loios and other neighbourhoods. Débora Soares, for instance, was Secretary of the African Association. The title is derived from Martin Heidegger's *An Introduction to Metaphysics*: 'Why are there essents rather than nothing? That is the question. Clearly it is no ordinary question' (Heidegger, 1959: 1). The English translation by Ralph Manheim states in a footnote that essents means existents, things that are, as a fictive present participle of the Latin *sum* (I am) (Heidegger, 1959: ix). I have some difficulty with Heidegger, and with connecting his enigmatic and possibly mystifying question to the content of the project, but a commentary by sociologist David Santos suggests that apart from a play on the name *Capital do nada* the idea of nothing has a power to question what is; and by extension the label of nothing linked to a blank on the city map has the power to question the marginalisation of that district and 'a community who, with great sacrifice, is still faithful to an area that is increasingly more desertified (*desertificado*)' (Santos, 2002: 160).

In January 2001 de Carvalho made visits to Marvila to meet people and take video footage, selecting twelve participants with whom to co-produce public images. They were Claudia Andreiea, Augusto Martins, Felisberto Tavares, Vanessa Carvalho, Lelio, Cassimir Carduso de Freitas, Débora Soares, Beto, Mário Moreva, Luis Lourenço, Francisco Fernandes and Maria Alexandra. The format in each case was a three-quarter length, standing photograph taken in front of a white backdrop, with the project title and a dedicated cell-phone number with which each participant was provided for the duration of the project printed beside it. When called, the participants described their leisure and work activities, such as Mário's graffiti, Beto's music, Maria's assistance to the Friar of the local church in Condado, and Sr Casimiro's quoits. The images were used in postcards, adverts in fashion magazines, posters at bus stops and on advertising columns, and a computer-sprayed image in one case, Débora, on the side of her block in Armador (fig. 8.1). Santos remarks that the technique resembles that of media promotions (Santos, 2002: 166).

Figure 8.1
*José Maças de
Carvalho,* porque
é existe o ser em
vez do nada?,
Marvila, 2001
(photo M. Miles)

Art-Ethnography

Citing Hal Foster's essay 'The Artist as Ethnographer' (Foster, 1996: 127–170) Santos remarks that, while practices in which contemporary artists adopt quasi-ethnographic methods in non-gallery projects contest the institutions of art, and received (often exclusionary) definitions of artist and community, the basis of association changes here to include 'the cultural and/or ethnic "other" in whose name the committed artist most often struggles' (Foster, 1996: 173, cited in Santos, 2002: 160–169). I return to Foster below, comparing his essay to remarks by Irit Rogoff (2000). Before that I ask how the project contributes to identity formation, bearing in mind the sociological discussion above.

There are at least three issues: authorship and voice; taxonomy and otherness; and tactics. To take the latter first, the tactics of Campino's project were surprise and provocation, emphasising the otherness of high culture as banal spectacle without seeking to widen access to it as though those present had no culture of their own. In contrast, de Carvalho worked in a participatory way with a long lead-in period and a collaborative process of disseminating the images produced. These were introduced in ordinary situations where they would be consumed with the more standard attributes of identity formation in consumer culture. The two projects appear complementary: culture-shock and culture-consolation, perhaps. This is not to say that the images and cell-phone conversations (the performative aspect being possibly the more important but without visual trace) were consolations for anything lost. Indeed, what they most strongly indicate is that people in Marvila have a culture (or cultures) and are as linked into mass communications as anyone else. To take the question of voice, while de Carvalho initiated and facilitated the project, the technique of photography means that the images were necessarily co-produced by those who stood in front of the lens, and could not have been made without their agency. The *Capital do nada* publication states:

> Equidistant from art and photography, this project consisted of a relatively long and phased process which offered visibility to those the artist defined as the 'heroes' of Marvila . . . Since the images were scattered around the country and city of Lisbon, many people called the heroes asking for information about them, the district and the event.
>
> (Extra]muros[, 2002: 152)

But why should the photographer need to retain the power of a lens? Handing out cameras for participants to use themselves is a standard community-arts tactic, and was used in other projects in *Capital do nada*. One answer is that specialist equipment produced high-level resolution images, just as they were professionally printed. Anything less would downgrade the worth of the image and by allusion the subject. Professionals, too, can facilitate a project by attracting funding and support, and by bringing it to other publics. The project is about visibility and extends the visibility of public exhibition to participants. Callers heard the uninterpreted voices of participants, not a representation of their stories. Santos cites Walter Benjamin's paper 'The Author as Producer' (Benjamin, 1998: 85–103) as a 'conciliation' (*conciliaçao*) between the artist and those regarded as social others, which not only delivers a commitment in a Marxist sense to social justice but also means that the desire for social change 'is transported to the domain of the awareness of the "ethnic other" resulting from the very racial and cultural pluralism of the outskirts [of the city]' (Santos, 2002: 163–164).

Power (over- or to?)

The above conciliation responds to the issue of taxonomies, in which those of production and reception, and citizen and other, are collapsed temporarily. For Walter Benjamin co-authoring was a means to insert artworks 'into the context of living social relations' (Benjamin, 1998: 87). But does such work shift power relations? Santos reads de Carvalho's project as evidence of a new art practice: 'after the situationist action of some artist groups of the seventies and eighties . . . another type of artist seems to have taken the place of the utopian artist . . . of the 20th century, the "artist as an ethnographer"' (Santos, 2002: 161). This raises several unresolved problems in ethnography (Hammersley and Atkinson, 1995; Hobbs and May, 1993), not least that the arrival of the ethnographer changes the situation to be studied through participant observation. A deeper question is that of objectification, or the relation of observing ethnographer-subject to observed ethnographic-subject. This adds to a difficulty in the social sciences that the scripts are given greater weight, so to speak, than the actors. On this, Tim May cites Dorothy Smith's *The Everyday World as Problematic: A Feminist Sociology*:

> The traditional methods of sociology objectify the social process, eliminating . . . the presence of subjects as active in bringing a concerted social world into being. The relations of people's real lives have been conceived as formal conceptual relations between factors or variables, expressing properties of social objects. These objects themselves have been elaborated as the constructs of sociological discourse embedded in its texts.
>
> (Smith, 1988: 152, in May, 1993: 72)

Ethnographic or socially committed art remains, nonetheless, distinct from the use of art by state agencies to address or conceal difficulties such as social exclusion produced by other areas of state policy. I do not want to replicate claims made by modernist avant-gardes that they have a special insight into life or social organisation, but I see projects like those discussed here as playing an important role not only locally but also in terms of broadening the issue beyond identity consumption.

An argument for artist-led projects is that they may take a dissenting stance towards the expediency described by George Yúdice:

> Advocacy for the centrality of culture in solving social problems is not new, but it took different forms in the past, such as the ideological (re)production of proper citizens . . . Although there have long been art therapy programs for the mentally ill and for the incarcerated, culture more generally was not regarded as a proper therapy for such social dysfunctions as racism and genocide.
>
> (Yúdice, 2003: 11)

Artists' groups such as Hewitt + Jordan (see case study 5.3) have used arts funding to question its intentions. But the ethnographic turn in art may result from envy, as Foster argues:

> Recently the old artist envy among anthropologists has turned the other way: a new ethnographer envy consumes many artists and critics. If anthropologists wanted to exploit the textual model in cultural interpretation, these artists and critics aspire to fieldwork in which theory and practice seem to be reconciled. Often they draw indirectly on the basic principles of participant observation.
>
> (Foster, 1996: 181)

Case study 8.3

Figure 8.2 Social Work and Research Centre, Tilonia, Rajasthan, India: puppet-making (photo M. Miles)

Comment

The Barefoot College at Tilonia, a village in central Rajasthan, is a Gandhian project established to work with the very poor. In India these include villagers, who work land owned by others at a subsistence level, under a weight of debts run up in times of poor harvest or drought. I visited the Barefoot College, or Social Work and Research Centre (SWRC), in 2003 and came away realising that as an academic I had nothing to offer except the skill to write about the experience. Most of the fifty staff are paid the legal minimum wage, and a monthly meeting of all staff has a power to adjust salaries according to circumstance (as when someone was thought to smoke too much). The College accommodates up to 250 villagers from the surrounding area, and a few from further away, for residential education over several weeks. Its work covers water harvesting – which it was set up to develop in 1972 – and agriculture, craftwork (for which women are paid money into bank accounts of which they are the signatories, not their husbands), health and maternity care (which has significantly reduced infant mortality and hence the need for large families), training in literacy and internet use for village women, the manufacture and maintenance of solar lamps and cookers (so that trees are not cut down for firewood), the building of houses for villagers in local materials, night schools for children and young adults (where they learn reading, writing, mathematics and about society, as I was told, when they cannot go to the government day school because they work in the fields in the daytime) and puppet making. The puppet shows do not follow traditional Rajasthani models in which the characters are kings and queens or nobles, but draw on everyday life for a cast of men, women, children, traders, farmers and government officials. The protagonist is Uncle. A show typically begins by Uncle asking the audience to guess his age (into three figures – I won't give it away). This lends him the status of an elder, to say unexpected things on agricultural or heath practices. And he can mock the government inspector who instructs that a well should be dug in a place that villagers know is dry. Uncle is a bit like the Joker in Augusto Boal's theatre for the oppressed, able to speak for the audience in ways they might not in a traditional society, 'a contemporary and neighbour of the spectator . . . [with] a protagonic function' (Boal, 2000: 177). Uncle is less formalised, a familiar character who turns up everywhere but carries the authority of someone who knows what he is talking about and needs no other authority to gain attention. Uncle's power is to utter, not to determine, and reminds villagers of their own knowledges in the way Paolo Freire (1972) asked participants in his adult literacy classes to begin with their own stories, not those of the dominant society and its cultural attributes and embellishments of hegemony.

In this they will lack adequate training. Returning to Foster's essay (which cites Benjamin's 'The Author as Producer'), he notes a danger that the artist may be separated from the subject of art-ethnographic work:

> the danger of ideological patronage is no less for the artist identified as other than for the author identified as proletarian. In fact this danger may deepen then, for the artist may be asked to assume the roles of native and informant as well as ethnographer. In short, identity is not the same as identification.
>
> (Foster, 1996: 174)

But Foster turns the question round so that the observer – in a tradition in which the observed are outside the dominant society and observed in isolation from it – and the observed are in the same world where 'the assumption of a pure outside is almost impossible' (Foster, 1996: 174).

Foster assumes an inside and an outside as defined, metaphorical and perhaps lost spaces. Rogoff's comment on the rhetorical question 'Where do I belong?' is more helpful:

> My inquiry does not attempt to answer the question of a location for belonging; it is by no means prescriptive since I have no idea where anyone belongs, least of all myself. It is, however, an attempt to take issue with the very question of belonging, with its naturalization as a set of political realities, epistemic structures and signifying systems. Thus it views questions of belonging through the cultural and epistemic signifying processes which are their manifestations in language.
>
> (Rogoff, 2000: 15)

Rogoff adds that part of the work to be done is 'that which mediates between the concrete and material and the psychic conditions and metaphorical articulations of relations between subjects, places and spaces' (Rogoff, 2000: 16). This is an interesting position from which to look at projects such as *porque é existe o ser em vez do nada?* The participants were not offered and did not seek belonging any more than they sought or were offered culture, already having culture just as they had everyday lives. Neither did the artist seek to cross over from identification to a shared identity. The question, then, is not whether art facilitates identity formation, which takes place without it, but whether art contributes to recognition of the fact that it takes place outside affluent consumption. In this vein John Roberts writes that much political art fails because it 'assumes that those whom the art work is destined for . . . need art as much as they need Ideas in order to understand capitalism and class society' (Roberts, 2001: 6).

Perhaps the crucial element of the projects in *Capital do nada* is that they attempted to recognise rather than donate the cultures of Marvila. There are complexities: Ian Toon argues that a distancing from the local 'is part of an important struggle for self-definition for . . . teenagers' (Toon, 2000: 144), which may affirm the attraction of fast food, brands and mass entertainment, though the availability of lifestyle commodities is limited for people on low incomes living in peri-urban and transitional zones. Toon argues that it is 'not only being seen but also seeing (the constant monitoring of self and other) that is an important practice of self-definition' in young people's sociation, 'where they are simultaneously actors and audience, consuming social space and each other' (Toon, 2000: 146). Similarly, Bonnie Leadbeater and Niobe Way comment on young black women's stories:

> Identity development is for all adolescents a process of self-evaluation that involves weighing one's own goals, ideals, beliefs, strengths, and talents against future visions of what one could be. It also includes comparison of oneself with the standards and ideals of others in one's community.
>
> (Leadbeater and Way, 1996: 3)

This is Arendtian publicity, a growth to maturity through the perception of others in renegotiation of perceptions. Perhaps, then, *porque é existe o ser em vez do nada?* is a means to publicity, using the vehicles of a more mundane publicity within a context which is politically active and has its own self-organising networks. The added-value, as it were, of art is that it lends these networks a visibility of which they are deprived by the dominant culture. Marvila's current publicity material, published in English (the global language) as well as Portuguese, tends towards a symbolic economy in stating 'Today Marvila offers many opportunities of development' but it also states 'Its energy comes from a dynamic social life' (Marvila, nd). Santos concludes that Carvalho reconciles the positions of Benjamin and Foster: 'The *Heroes of Marvila* . . . interrogate the mediation of subjects' self-knowledge by others, and knowledge of others by themselves . . . [in] a complex process of identity interaction' (Santos, 2002: 166, my translation).

Conclusion

I would relate the projects in *Capital do nada* to Toon's interest in ambivalent spaces in young people's occupation of urban spaces:

> The public spaces of the street, park and shopping centre are important in the social process of style creation because they provide a context in which the performance of appearance takes place. For example, these teenagers use their looks as distinctive elements in the exploration and expression of identities which they enact spatially through modes of self-display.
>
> (Toon, 2000: 145)

In Marvila there are social spaces in the blocks and community centres which take the place of streets, parks and the mall. While Toon argues that shopping spaces are transformed as arenas for display, it could be that Marvila is, too, a mundane environment open to playful transformation. That seems a fitting conclusion which I would qualify only by saying I make no special claim for the projects discussed beyond that they open questions of socially grounded cultural experience, using methods which are appropriately collaborative and at times provocative. As Catherine Belsey says, culture is 'the inscription in stories, rituals, customs, objects, and practices of the meanings in circulation at a specific time and place' (Belsey, 2002: 113).

9 Permeable cultures

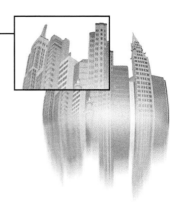

Learning objectives

- To draw a small number of contingent inferences from the material of preceding chapters
- To consider the permeability of urban cultural forms through the case of the urban grid
- To reconsider the distinctiveness of cities and cultures as mutually permeating concepts
- To ask if the conditions of urban cultures are conducive to a right to the city

Introduction

This has been a difficult book to write because it covers a large field in which the concepts *culture* and *city* are subject to continuing redefinition. I end with a feeling that concepts are part of the problem, though I do not advocate an approach which is entirely empirical. In a creative tension between ideas and evidence, then, I embark on this final chapter concerning the permeability of cultural formations. This is not exclusively a dimension of urban cultures but is particularly evident in the specifically urban form of a street plan such as the familiar grid. The recurrence of the grid in different societies and histories does not guarantee a uniformity of meaning, however. It is easy to assume that grids are inscriptions in the manner of the engineer in Descartes's *Discourse* drawing regular places according to his fancy (Lacour, 1996), but this is an idea specific to European modernity. I begin the chapter with an exploration of the urban grid to find diverse and conflicting meanings in its specific appearances, looking particularly at the form of cities in the Americas as discussed by Setha Low in

On the Plaza (2000). The grid, still, implies an ordering of space according to a plan. In face of increasing recognition today that cities are sites of mutability and diversity I ask if such an ordering is sustainable. Finally, I ask how I can understand from this and preceding sections of the book what is meant by a right to the city (in Henri Lefebvre's terms) and how it might include the imagination and realisation of future urban conditions. In case studies I cite Michel Tournier's reworking of the Robinson Crusoe story in *Friday*; and the work of the Sarai Media Lab, Delhi. Finally, I illustrate a bunch of carnations placed on the Soviet War Memorial in Berlin in 2005.

The grid

The Americas

Architects Ana Betancour and Peter Hasdell write of the founding of Buenos Aires in 1536 that Don Pedro Mendoza first drew the city as 'a colonial grid surrounded by written script on all four sides' (Betancour and Hasdell, 2000: 155). It was a rhetorical device emblematic of 'the gap between the inner landscape, an imagined utopia of riches and civilisation, and a city that did not yet exist' (Betancour and Hasdell, 2000: 155). Mendoza died of syphilis – Europe's gift to the Americas as to so many colonised places – and the city fell into ruin to be refounded in 1580 – again according to a grid (drawn on cow-hide). Betancour and Hasdell cite Angel Rama's argument in *The Lettered City* (1996) that the colonial city of the Americas is prefigured in literary and legal forms such as the Law of the Indies, which constructs the conquered society as wild nature in face of European literacy. But the relation is more ambivalent. If as Betancour and Hasdell write, 'The implementation in the colonies of the Spanish cuadricula, and the rational urban grid from the late 19th century are very much alike' (Betancour and Hasdell, 2000: 156), it could be assumed that the gridiron plan of New York (1811) is an ordering of land in which settlement represents civilisation and the land untamed nature. This follows Pierre L'Enfant's plan for Washington and William Penn's for Philadelphia, both of which represent an Enlightenment ordering of space according to the perceived laws of natural growth, but the New York grid was a parcelling of land as lots of equal size for speculative purchase (Lynch, 1981: 180). Grahame Shane, citing Manfredo Tafuri (1976), reads the New York grid plan as 'symbolic of the pragmatic, machine-age mentality of the Enlightenment land surveyors (who also subdivided the American continent into gridded lots)' (Shane, 2002: 232). Matthew Gandy reads it as making 'the pragmatic underpinnings of capitalist urban design far more explicit than in . . . European counterparts' (Gandy, 2002: 11). The insertion of Central Park, opened in the winter of 1859, interrupted the grid by introducing a renaturalised landscape, and Gandy contrasts the 'Promethean obliteration of the "first nature" of Manhattan Island' in the grid with Central Park's Arcadian vision of 'an imaginary nature that in no way contradicted the fundamental dynamics of nineteenth-century urbanization' (Gandy, 2002: 110). For the Conquistadors, the grid offered spaces of display, or power, while the New York gridiron's relation to capital sets in a related but not identical framework. Further investigation reveals other foundations for a grid, and that the form may not be simply a European inscription.

Greece

New York's 'checkerboard of individual land parcels awaiting investment' (Gandy, 2002: 81) contrasts with the grid of Greek colonial cities. From a philological investigation of the Anaximander Fragment, also interpreted but differently by Heidegger (1984: 13–58), Indra Kagis McEwen (1993) links the Greek grid to the practicalities of material culture. She rejects Irad Malkin's (1987) hypothesis that the Greek city is planned – its materialisation simply realising a preconstructed concept – and that colonisation is a basis of the *polis*; instead McEwen argues that the regularity of streets in the Greek orthogonal plan has 'much more to do with the notion of allowing *kosmos* to appear through their rhythm than with planning in the modern sense of the word' (McEwen, 1993: 80). She continues that the orthogonal plan *appears* just as the threads of a well-made cloth, woven perpendicularly, articulate a cosmic order; and draws a further analogy with the perpendicular joints of timbers in a well-made ship. Appearance is a setting apart from the ordinary, constructed as performance in keeping with an order which is all-pervasive but brought to visibility in certain things. She adds that colonial cities influenced mother cities as sailors became settlers: 'Between the *metropolis*, or mother city, and the new foundation, the city existed as a ship' (McEwen, 1993: 80).

Spain and New Spain/Mexico

The grid with central square (*plaza mayor*) is typical of cities in Spain and its conquered territories. Setha Low examines the development of this form in sixteenth- to seventeenth-century Spain and Latin America. The navigator Christopher Columbus signed the articles of agreement (*capitulaciones*) for his voyage at the garrison city of Santa Fe in Granada in April 1492, a year after its completion in a final stage of the ethnic cleansing of Spain by its Catholic monarchs known as the expulsion of the Moors. The form of Santa Fe was the basis of the plan of Santo Domingo, founded by Bartolomé Columbus in 1502 (Low, 2000: 95). George Foster (1960) traces the form of *bastides* (fortified hill-towns) in Navarre and Languedoc to Santa Fe. Both contrast in formality to the cumulative cities of medieval Europe and the Islamic cities of southern Spain and north Africa (though Low mentions Islamic gardens as a possible influence – 2000: 95). Low continues that the case for a European origin for Spanish colonial grid-plan settlements is drawn from the Law of the Indies written by Philip II from 1509 onwards and published in a unified form in 1537; but the evidence of sixteenth-century narratives offers a parallel history to that of the Laws.

According to his account, *De insulis inventis*, published in Basle in 1493 Columbus used the timbers of the wrecked Santa María to build a fortress he called La Navidad on the coast of Hispaniola. According to Low, the Spanish encountered indigenous towns there in which 'The streets were generally straight, and plazas were built for ceremonial use . . . a definite Taino [indigenous] building design in which, after felling the trees to clear a place for the plaza, four "streets" were hewn out in the form of a cross' (Low, 2000: 96). Excavations show rectangular spaces enclosed by flat stones. A text by Fray Bartolomé de las Casas notes open squares in front of the houses of an elite. Low notes that his work has been 'critiqued as a revisionist history to empower the native populations' (Low, 2000: 97) but regards his evidence as useful. A new city for 2,500 settlers in 1502–3 used a geometric plan drawn up by the new Governor Fray Nicolás de Ovando, who 'coordinated selected urban sites, controlled municipal appointments and determined the disposition of lots around the plaza' (Morse, 1987: 171, cited in Low, 2000: 97). Gonzalez Fernández Oviedo y Valdés compares San Domingo favourably to Barcelona:

> Since the city was founded in our own time, there was opportunity to plan the whole thing from the beginning. It was laid out with ruler and compass, with all streets being carefully measured.
>
> (cited in Low, 2000: 98)

This is quintessentially modern in separating plan from execution, and makes an interesting comparison to René Descartes's use of an architectural metaphor:

> Thus one sees that buildings which a single architect has undertaken and completed are usually more beautiful and better ordered than those which several architects have attempted to rework, using old walls that had been built for other ends. Thus these ancient cities, . . . are normally so poorly proportioned, compared with the well-ordered towns and public squares that an engineer traces on a vacant plan according to his free imaginings.
>
> (Descartes 1963 [1637] : vol. I, 579, in Lacour, 1996: 33)

Descartes is thought to refer to the new cities of Nancy and Charleville, but the material cited above predates these new cities by around a century.

To return to New Spain, Low concludes that San Domingo's grid plan reflects both Santa Fe and the plaza of Islamic Córdoba (Low, 2000: 99). She goes on to demonstrate an indigenous source as well. Cities such as Tenochtitlan and Cuzco were centred on squares surrounded by temples, and the regularity of these sites was admired by the Spanish in contrast to the medieval cities they left behind. Tenochtitlan was a city of around 200,000 people, possibly the largest in the world, completed in 1325 on a marshy site. Its straight causeways led to a central plaza representing a cosmology structured by four cardinal points. Nahua cities, too, were rectilinear, the central plaza being used for ceremonial purposes and as a market. Low cites research showing that social status was measured by the distance of a dweller's house from the plaza, and writes, 'The Mexica were developing principles of general planning in order to achieve an efficient urban organization' (Low, 2000: 107, citing Calnek, 1978; Zucker, 1959). Similarities of socio-economic organisation enabled the Spanish to govern with relative ease, while the rebuilding of Mexico City on the ruins of Tenochtitlan incorporated elements of indigenous urban planning later informing practices in Spain. Low cites contemporary sources on the grandeur of Tenochtitlan:

> Some of our soldiers who have been in many parts of the world, in Constantinople, in Rome . . . said they have never seen a market (plaza) so well laid out.
>
> (Diaz del Castillo, 1963: 235, in Low, 2000: 113)

and, from Cortés:

> The city itself is as big as Seville or Cordoba. The main streets are very wide and very straight . . . This city has many squares where trading is done and markets are held continuously. There is also one square twice as big as that of Salamanca, with arcades all around.
>
> (Cortés, in Low, 2000: 114)

Tenochtitlan was destroyed after a siege of 75 days. In 1521, changing his mind about rebuilding the city on a new site, Cortés ordered its rebuilding *in situ*; this followed the Catholic reoccupation of Moorish cities in Spain after the expulsion, and in the plan drawn in 1523–24 the plaza remained the central feature. The grid with central plaza was, then, an established Mesoamerican urban type, while the Spanish invasion repeated the strategies of that of Moorish Spain – completed as the first voyages were planned.

Low concludes, refuting an inference from the Laws of the Indies that post-conquest grid cities are colonial inscriptions, that the Plaza Mayor in Madrid (1619) is 'a response to the beauty of the designed plazas of the Spanish American

Case study 9.1

From Michel Tournier, *Friday*

But Robinson could not wholly regain his human dignity except by providing himself with a better dwelling than a cave or screen of leaves. Now that he had as companion the most *domesticated* of all animals [a dog], he needed to build himself a proper setting for domesticity, a *domus*, a house.

He decided to build it near the entrance to the cave where all his wealth was stored, which was also at the highest point of the island. He began by digging a wide, rectangular trench some three feet deep, which he filled with branches covered with a deep layer of sand. On this well-drained foundation he set up walls of palm logs, notched and fitted together, the interstices being filled by their scaly bark with its growth of vegetable matting. He constructed a double-sided roof of light poles plaited with reed which he covered with the leaves of rubber plants, using them like slates. Then he plastered the outside of the walls with a mortar made of clay and chopped straw, and completed his sand floor with a rough paving of flat stones, over which he spread goatskins and straw mats. A few articles of furniture were made of osier, the dishes and lanterns he retrieved from the *Virginia*, the spyglass, the cutlass, and one of the muskets hanging on the wall created an atmosphere of comfort and even intimacy in which he constantly rejoiced. Seen from outside, the first dwelling had the surprising look of a tropical log cabin, crude but neat, with a fragile roof but massive walls, which seemed to Robinson a mirror of the paradox in his own situation. He was conscious both of its practical inadequacy and of its symbolic, and above all moral, importance . . . By degrees the cabin became for him a sort of museum of civilised living, and he never entered it without a certain sense of ritual.

(Tournier, 1997 [1967]: 64–65)

Comment

Tournier reworks the story of Robinson Crusoe as a saga of civilisation's rebuilding in the site of its absence on an island off the coast of Chile after the catastrophe of shipwreck. The separation of culture as the arts and culture as a way of life is collapsed by Robinson's sudden fall, as he perceives it, into a state of necessity in which he is the only human survivor. After a while he finds the ship's dog, Tenn, but the rest of the crew's bodies are taken by the sea. He lives like a non-human animal, making do, wallowing in a swamp and forgetting the routines of human society; then builds a representation of his previous, civilised life in the hut described above. The improvised shelter takes on layers of value as Robinson's lifestyle becomes more, as it were, re-Europeanised. He begins to wear clothes salvaged from the ship, and finds casks of explosive, grain and a quantity of tobacco. Having built structures which combine defence with domesticity he takes to dressing in formal attire – jacket, breeches, stockings and shoes – before entering the space of such ordering. As Tournier writes, it is as if Robinson were paying a formal visit to himself (or what he perceived as best in himself, to be kept in reserve).

Robinson's artificial way of life is his means of transforming his circumstances and hence his material life as well. In his state of crisis the distinction between culture as agriculture, as a way of life, and as the development of a cultivated perception of the world collapses. Incrementally the cultivation of the island and himself, the island's and his own domestication, lead to a re-emergence of civilisation. He produces a surplus in his agriculture, storing grain, and extends his architecture as a model of an early colonial built environment. He writes a journal, records the data of sowing and harvesting, arranges things in order and assumes himself superior – because he has this vocabulary – to Friday, an Araucanian who arrives on the island by boat with a group of men and a sorceress. He escapes when he is marked for sacrifice, but his escape is to a society of one other person for whom escape is not an option. Gradually a working relationship evolves between Robinson and Friday, which entails Friday's acceptance of Robinson's laws and language. Robinson, as a sign of trust, leads Friday to the boat he once made. Its timbers have disintegrated, and Tenn falls through the termite-eaten deck. 'The *Escape* was little more than a heap of sawdust' (Tournier, 1997 [1967]: 137). Friday laughs, and takes on most of the manual work. Robinson does not credit Friday with a culture in the way he attributes one to himself, now as Governor of the Island, drafting a set of regulations and punishing himself for transgressions. Friday steals some of Robinson's tobacco. The hut and grain store are destroyed in an explosion which also causes Tenn a fatal shock. A ship arrives, Robinson takes dinner with its captain, but when it sails it takes Friday while Robinson, and the ship's cabin boy, remain on the island now called Speranza.

New World' (Low, 2000: 94), and the Latin American plaza, 'syncretic in the sense that it is no longer Spanish or indigenous but, through a series of historical and sociopolitical processes, has become emblematic of Latin American culture' (Low, 2000: 122).

Agency

Inter-culturality?

I do not want to map the experience of the grid in Spain and Mexico on to all other cases, but infer that cultural formations can in some conditions permeate each other; and the relation is not only from dominant to dominated. Margaret Thompson Drewal notes, for instance, that white, North American culture has changed by absorbing elements of the cultures of groups such as ex-slaves. She argues: 'To win the transcultural war is to transform dominant culture's very identity . . . hegemony persists, but as Raymond Williams observed, the internal structures of hegemony have to be renewed, recreated, and defended' (Drewal, 2000: 116, citing Williams, 1980: 38, 45). Under a regime of globalised capital, too, shifts occur reflecting spontaneous reactions by consumers, such as the growth of a market for organic food among affluent groups, though the market is adept at subsuming more or less any departure. Steven Feld sees the growth of world music serving market expansion through 'artistic encounters with indigeneity' in a 'consumer-friendly multiculturalism' (Feld, 2001: 213, citing Kumar, 1995; Harvey, 1989; see also Abu-Lughod, 1991).

John Tomlinson argues that globalisation is a globalisation of Western culture and consuming values but not a coherent development in that Western patterns of consumption co-exist with non-Western attitudes to gender and sexuality. Tomlinson adds, citing Arjun Appadurai (1990), that 'Movement . . . always involves translation, mutation and adaptation as the "receiving culture" brings its own cultural resources to bear' (Tomlinson, 1999: 24). The impact of global-isation is ambivalent, as in a capacity of communications technologies 'to enable and to disempower' (Tomlinson, 1999: 26), while anthropology has abandoned an assumption that 'other cultures were there, like the flora and fauna of the natural world, to be catalogued . . . but there was no thought they could ever *themselves* engage in the practice' (Tomlinson, 1999: 27). Jennifer Robinson argues that 'the time chart of urban theory neatly flips around, from setting urban experiences in places such as Africa and South America into the West's past' when 'conditions . . . seen as undeveloped are indicators of the futures that cities everywhere might face' (Robinson, 2006: 91), citing Lagos as related by the Dutch architect Rem Koolhaas. Tomlinson concludes that globalisation does not represent a Western 'cultural grip' regardless of the West's economic power, and that 'the global future is much more radically open than the discourses of homogenisation and cultural imperialism suggest' (Tomlinson, 1999: 29).

Yet globalisation centralises economic power even if it allows culture to mutate, and as Lefebvre observes:

> Throughout history centralities have always eventually disappeared – some displaced, some exploded, some subverted. They have perished sometimes on account of their excesses – through 'saturation' – and sometimes on account of their shortcomings, the chief among which, the tendency to expel dissident elements, has a backlash effect. Not that these factors are mutually exclusive – as witness ancient Rome, which suffered both internal saturation and the assaults of peripheral forces.
>
> (Lefebvre, 1991: 333)

The expulsion of dissidents tends to follow a suspicion of strangers, as in the Venetian ghetto (Sennett, 1995: 220–240) and the War on Terror, but in the latter case does not lead to a reduction in ethnic restaurants in cosmopolitan cities. As signs are disconnected from the detail of signifieds, it may be that cultural cross-currents have a role in affirming (by normalising) difference, and destabilising dominance by asserting mutability. This is not to deny economic reality but to enlarge it, as Peter Worsley does in arguing that 'social science is wider than political economy; wider, too, than a purely structural sociology devoid of culture' (Worsley, 1999b: 37). I agree, and with the idea that there are cultures of economics and politics, and economic and political determinants of cultural production and distribution.

Reclamations

Seeking to reclaim the idea of multiculturalism from its negative reception, Leonie Sandercock proposes an intercultural co-existence as a way to address difference in 'mongrel cities', listing its components as 'identity/difference; the centrality of conflict, or an agonistic democratic politics; the right to difference; the right to the city; and a shared commitment to political community' (Sandercock, 2006: 46, citing Mouffe, 2000; Parekh, 2000; Tully, 1995; Connolly, 1991). One strand in such a contingent urban cultural and political future may be what Yves Fremion calls an orgasm of history (2002) in which multiple publics organise their own lives and interpretations of the world to inform those lives:

> Self-management is when individuals take their own fate and future prospects into their own hands, without any specialist intermediary, and it is as old as the world itself; except that in the 20th century we have . . . an acute awareness of it.
>
> (Fremion, 2002: xii)

Case study 9.2

From Nitin Govil, 'The Metropolis and Mental Strife'

The city science fiction film takes full advantage of recoding familiar architectural imagery in a transformative fashion. The iconography of Los Angeles
is prominent in, for example, *The Terminator*, whose 'tech-noir' aesthetic
illustrates SF's fascination with the city via the classic American detective films
of the 1940s and '50s. Similarly, *Bladerunner* is a postmodern neo-noir that
brilliantly inverts Disney's theme-park fascination with the topography and
urban psychology of LA. In contemporary American cinema, LA displaces New
York as the centre of science fiction's urban imaginary in films like *Predator 2*,
Demolition Man, and *Terminator 2*. Los Angeles also figures strongly in recent
SF films about virtual reality, which demonstrate the powerful influence of
cyberpunk as a major urban aesthetic. However, particularly since the early
1980s, New York iconography has become prominent in SF dystopias that
range from *Escape from New York* to *Batman* and *Dick Tracy* . . .

Recent millennial and apocalyptic anxiety has, however, restored New York
fully to the forefront of the mainstream of SF film. Los Angeles figures
prominently in the popular imagination of natural disaster, subsuming anti-
environmental and man-made profit-oriented design under the narrative logic
of ecological revenge, and New York City has been destroyed in classic SF
films like *Deluge*, *When Worlds Collide*, and *The Beast from 20,000 Fathoms*.
Yet there is a distinctive iconographic reversal in these new, mega-budget,
SF films . . . the graphic of repeated destruction . . . urban annihilation . . . in
Deep Impact the World Trade Centre towers topple under the force of a tidal
wave that destroys all of New York City.

(Govil, 2002: 82)

Comment

This extract is from *Sarai Reader* (available on-line at www.sarai.net). Sarai Media
Lab, which produces this annual anthology of urban writing, is a self-organised channel
of global communication and critique. Rajeev Bhargava writes in the same edition:

One such message [of '9/11'] which the poor, the powerless and the culturally
marginalised would always like to have communicated to the rich, powerful
and the culturally dominant, although not in this beastly manner, is this: we
have grasped that any injustice done to us is erased before it is seen or spoken

about, that in the current international social order, we count for very little; our ways of life are hopelessly marginalised, our lives utterly valueless. Even middle-class Indians with cosmopolitan aspirations became painfully aware of this when a countrywide list of missing or dead persons was flashed on an international news channel: hundreds of Britons, scores of Japanese, some Germans, three Australians, two Italians, one Swede. A few buttons away, a South Asian channel listed names of several hundred missing or dead Indians, while another flashed the names of thousands with messages of their safety to relatives back home.

(Bhargava, 2002: 200–201)

Sarai Media Lab is based in Delhi, a collective of around twenty people operating through the internet and using free software tested and adapted continuously by users. The name Sarai is derived from sarai (or caravanserai), a walled enclosure for travellers in a city erected in charity to provide shelter, food and companionship for strangers. Sarai is housed in the Centre for the Study of Developing Societies, where it holds weekly seminars, hosts on-line inter-media creative projects, undertakes academic research on urban questions, engages with the dissemination and development of free computer software and has a digital archive. Among current research topics are: the impact of new media on Hindi popular culture; urban ecologies and governance; the uses of free software in experimental communications media; and new spatial practices of work and dwelling. Sarai describes itself as constructing a space

where old and new forms of media, their practitioners, and those who reflect on, or critically examine these practices, can find a convivial atmosphere, and enter into shared pursuits that will create a renewal of public cultures within and across city spaces.

(*Sarai Reader*, 1, p. v)

The introduction to Sarai's first Reader describes a 'real and imagined' public domain constituted as an alternative to that of power: 'This republic without territory is a sovereign entity that comes into being whenever people gather and begin to communicate, using whatever means that they have at hand' (ibid.). It has neither borders nor officials, its 'ministries and embassies' located 'throughout the world in cafes and public libraries, in the lobbies of cinemas, in theatres, in obstinately independent radio stations, in websites, between the pages of free zines. The source code of its constitution, written in free software, is open to all to amend' (ibid.). Among the aims of *Sarai Reader* are to bring marginalised and erased voices into audibility, and to foster forms which may seem chaotic or contradictory in street culture extend 'into eccentric orbits of freedom and solidarity' (*Sarai Reader*, 1, p. vi).

This leaves the problem of how such consciousness translates into action, and for me in this book how, given that more or less all major political and social upheavals take place in cities, cultural work operates to advance this scenario among multiple urban publics.

I have written elsewhere of interventionist tactics in collaborative art practices (Miles, 2004). Another possibility is to see the internet as enabling mass communication without intermediaries, and its networks of mobilisation as conducive to direct democracy. I have some sympathy for this, and for the idea proposed by Armin Medosch that 'The technical and two-way structure of the internet makes it possible to break out of the pattern of broadcast media' (Medosch, 2004: 143). Medosch argues that small groups constituting fragmented on-line publics can bring issues to public attention that are erased by the global news industry, which relates to Sandercock's call for an agonistic politics. This, too, is outside my scope (but see case study 9.2). Finally, then, I want to look in a more theoretical way at Sandercock's fourth element, the right to the city – a concept from Lefebvre.

Right to the city

For Sandercock the right to the city means that 'negotiating peaceful intercultural coexistence, block by block, neighbourhood by neighbourhood, will become a central preoccupation of citizens as well as urban professionals and politicians' (Sandercock, 2006: 48). Her phrasing has the feeling of a peacekeeping operation. The aim is to assert a right to occupy the city's neighbourhoods and public spaces, and its public sphere of determination. Sandercock acknowledges Ash Amin's idea (2002) of a politics of liveability and James Donald's (1999) awareness of migration as a key factor in urban futures, with Iris Marion Young's concept of group difference (1990); but Lefebvre is not listed in her references. Since he is a key source for me I take this absence as cause to reflect on his relevance now, when his work is situated in a discourse prior to multiculturalism and post-colonialism, and he uses a normative masculine – writing most of his key works between the 1950s and the early 1970s. He is engaged in a critical dialogue with French Marxism (see Shields, 1999: 89–94), since overtaken by the events of 1989 and 1991 (see Chapter 7), if not 1968. Yet the right to the city is a pervading idea in Lefebvre's thought, which, in a characteristically fugue-like style of writing, he does not limit to definition as a concept. I can see, given the frequency with which he is cited anyway, why Sandercock might see Lefebvre as outside the terrain of her inquiry, though for me his approach retains currency.

In 'The Right to the City' – first published as *Le Droit à la ville* in 1968 – Lefebvre begins from an analysis of need, arguing that social needs (rather than the conventional model of individual need) exhibit complementarities: security and opening, certainty and adventure, organisation and play and so forth. He claims that the need for creativity is ignored in urbanism, while one characteristic of what he reads as a failed millenarianism in planning regimes is a domination of the city by the countryside. This is contrary to early (German) sociology, and, while conditions such as proximity (see Chapter 1) tend to be identified as defining urban dwelling, Lefebvre looks instead to industry as the ground for a new praxis (Lefebvre, 1996: 147–150).

He then advances two concepts: *transduction* as a model of constant reflexivity in the relation of theoretical model and data; and *experimental utopia* as investigation of how utopian models of social organisation operate in practice, and their underlying assumptions. Observing the limits to a postwar rational planning approach, he states:

> Knowledge can therefore construct and propose models . . . [of] urban reality. Nevertheless, such a reality will never become manageable as a thing and will never become instrumental even for the most operational knowledge. Who would not hope that the city becomes again what it was – the act . . . of a complex thought?
>
> (Lefebvre, 1996: 154)

I have doubts that cities ever were that, or perceived as that, since the city as act rather than site is a late modern idea; and am unsure about the appeal to a non-specific past which tends to lapse into Arcadia or Eden. Lefebvre's rhetorical style hints at, however, and the text later confirms, a projection not backwards but forwards to revolt – in context of the alliance of students and workers in Paris in May 1968. Lefebvre says, 'It cannot but depend on the presence and action of the working class, the only one able to put an end to a segregation directed essentially against it. Only this class, as a class, can decisively contribute to the reconstruction of centrality destroyed by a strategy of segregation and found again in the menacing form of *centralities of decision-making*' (Lefebvre, 1996: 154).

After 1968 it would have seemed nostalgic to say this, and Herbert Marcuse, in the years following the failure of revolt in 1968, begins to turn to an aesthetic dimension, after uttering the idea of society as a work of art (1968: 185), and the traditional role of a European intelligentsia alongside an argument that new quasi-biological needs (for freedom) are produced in late capitalism. But, as Stuart Elden points out, Lefebvre was aware of the growth of new towns in France in the postwar period, and the deficiency of Marxism in dealing with issues such as urban housing (Elden, 2004: 141). Elden also notes that Lefebvre saw the division of elites and excluded groups within urban sites as replacing the traditional divide of city from countryside (Elden, 2004: 151). Elden's latter point indicates another

Case study 9.3

Figure 9.1 *Soviet War Memorial (1946), Tiergarten, Berlin, 2005 (photo M. Miles)*

Comment

I saw this unexpectedly, in Berlin for a European arts education conference and escaping for a couple of hours to walk through the city, taking a central axis from Charlottenburg to Unter den Linden (where I like Café Einstein). On a previous visit I saw a coach bearing the name of the operating company in large letters on its side: Marx. It was parked next to the Soviet War Memorial (Lev Kerbel, 1946), built in part with marble salvaged from Albert Speer's new Reichkanzlei, a large and formal construction in memory of the soldiers who died liberating the city in 1945. I walked into the monument's central space to see a bunch of fresh flowers placed on the horizontal slab beneath the central column. I have no idea who left it or why, but suppose it was in sincerity, perhaps in memory of an individual, perhaps in gratitude for events sixty years previously, or in a generalised recognition that there is an alternative ideology to that of market economics. The histories referenced – the defeat of Nazism, the Soviet period in the GDR or more tangentially German reunification after 1989 – are histories in which I did not participate and on which I have no claim of authoritative interpretation. As I suggest in Chapter 7, I see a trace of hope in the socialism which failed in the years between 1917 and 1989 to achieve its aims of a withering away of the state apparatus and emancipation from toil in appropriation of the means of production. I have no voice in the reconciliation or forgetting of the Soviet period. My visits to such monumental signs of that revolution tend to be for reasons I do not understand, and which do not include an attraction to the architectural style of the edifices or the statues which embellish them. I go, I suppose, from curiosity and perversity, a stranger. The flowers were red, the colour of the workers' flag.

aspect of meaning in Lefebvre's remarks – the urban revolution is, apart from political revolt, a revolution of settlement, departing radically from a model reflecting ruralism or incremental development of a city within a rural hinterland. The urban *is* the revolution (as condition), as he elabourates elsewhere (Lefebvre, 2003). Logically it is also the site of revolt (as political tactic); and in Lefebvre's reconsideration of the Commune and Haussmann's clearance of working-class quarters in Paris, in *La Proclamation de la commune* (published in French in 1965), a context for his emphasis on the working class becomes clearer (Elden, 2004: 153). These ideas were developed in discussions with the Situationists, and produced a theory of moments of liberation within the routines of everyday lives (see Shields, 1999: 58–64).

Returning to the text, Lefebvre asserts a need for urban reform unconstrained by the prevailing society; and for radically imaginative future thinking on urbanisation. He makes a case for art as creating intelligibility in specific ways; and argues that among present contradictions (in the 1960s) are 'the realities of society and the facts of civilization . . . genocide, and . . . medical and other interventions which enable a child to be saved or an agony prolonged' (Lefebvre, 1996: 157). He remarks that a right to nature is extant when urban dwellers take leisure in green spaces, which he sees as reminiscent of a rural framework, and proposes the right to the city as, in a way, its contrary: 'a transformed and renewed *right to urban life*' (Lefebvre, 1996: 158). Rather like a Fourierist vision of a harmony of artisanal and intellectual work, this revolves around the coincidence of art and science, and again the agency of the working class. I would say that the working class can (and must) be taken out of the equation, but that the core of Lefebvre's ideas remain both helpful and illuminating. Among other reasons for that I would say he enlivens, socialises and brings to real political possibility a utopianism which had traditionally (and in international modernism) been aestheticised or functionalised.

In *The Urban Revolution* (published in French in 1970 – after the events of 1968 and before *The Production of Space*), Lefebvre writes, extending the above, 'The concept of an urban society has freed itself from the myths and ideologies that bind it, whether they arise in agrarian stages of history and consciousness or in an unwarranted extension of the representations borrowed from the corporate sphere (industrial rationalism)' (Lefebvre, 2003: 165). As I reread Lefebvre some of his remarks resonate with contemporary debates on consumption and monopoly capitalism. Others – references to the working class – seem encapsulated in history. Yet there is a deep desire to transform utopian dream into practical urban programme in his work; and a use of the term *urban* more than *city* to denote a condition rather than a geographical site. I am left wondering if Sandercock's image of block-by-block negotiation of co-existence will in time be similarly encapsulated in a historical moment characterised by urban conflict, terror, surveillance and sports utility vehicles (SUVs) as well as by the primacy of difference in academic discourses. Having said that, her essay on mongrel cities (2006) sets out an agenda for urbanism which, with some creative extension, could be seminal in the present decade.

Non-conclusions

I have attempted in this chapter to give some indication of the complexities of current reconsiderations of urban form as a vehicle of ordering, and potentially of an imaginative reconceptualisation of cities as active processes. But I find myself defeated by the linguistic conventions in which I try to say that the kind of reconceptualisation I glimpse, in Lefebvre for instance, no longer has need of the model of a concept. Hyvrard's encept comes to mind (case study 8.1) but is difficult because it departs from the standard terminology which represents but also reinforces standard ways of thinking based on a separation of idea from execution (like the planned city). It is hard to say something in a language of which I have a fragmentary syntax and a partially understood vocabulary, and which exists only in experimental terms, which may sound rather obvious. My doubt, still, leads me to defer a conclusion to this chapter and the book as a whole. Were I to offer one it might reduce what I have written to a few bullet points, when I believe in the value of extended conversation; and would close the argument, while one of the inferences I take from reading Adorno is that, faced with polarities such as the social and aesthetic, Adorno's tactic is to keep the argument going at any price, often via a refusal of polarisation in favour of an axis of creative tension. Lefebvre does something coincidentally similar in the last paragraph of *Everyday Life in the Modern World* (written in Paris in 1967), citing a protagonist of the Jacobin period of the French Revolution:

> Saint-Just said that the concept of happiness was new to France and to the world in general; the same could be said of the concept of unhappiness, for to be aware of being unhappy presupposes that something else is possible, a different condition from the unhappy one. Perhaps today the conflict 'happiness-unhappiness' or 'awareness of a possible happiness-awareness of an actual unhappiness' has replaced the classical concept of Fate. And this may be the secret of our general *malaise*.
>
> (Lefebvre, 2000: 206)

Earlier in this text Lefebvre claims that the concept of revolution is still valid. As in *The Right to the City*, he argues for a confluence of economic, political and cultural revolution, and that 'a revolution cannot be other than *total*' (Lefebvre, 2000: 197). I find it is easy to be drawn into such convulsive remarks as a foil to the relative ordinariness of most of my academic work in an institution (of higher education). Yet the idea lingers that revolution is total, if unachieved in political programmes, because it is always incomplete. Cities are events and permanent cultural revolutions.

If, then, there is a defining quality of city life (as I inquired in Chapter 1) it is a positive incompleteness, produced in conditions of proximity but extending to a theoretical limit of the production of the city (necessarily ephemerally) by its inhabitants. In this, culture as the arts and media fuses with culture as a way of life to produce – again I am informed by Lefebvre here – a creative way of life. This is equivalent to the idea proposed by Joseph Beuys in the 1960s that everyone is an artist, and Marcuse's idea of a society as a work of art. In another sense, of course, as Jane Jacobs cautions (1961), a city is not a work of art but where people live and work. I accept the tension between that and the previous point as a research agenda. Perhaps it is not exactly the agenda of traditional cultural production in service of elites, or of modernism in service of the pursuit of autonomy mirroring the fantasy of a coherent, unified citizen-subject (see Belsey, 1985; 2001; 2002; 2005). Neither is it the agenda of the culture industries as I persist in calling them to reassert a critical perspective, countering the rhetoric of culturally led urban redevelopment (see Chapters 5 and 6). But it links as a research agenda situated across several academic discourses to the issues of empowerment and development discussed in Chapter 8, and is contextualised by both the complex cross-currents of a world in which the demise of an East bloc produces a former West as well (Chapter 7); and the mutabilities discussed over a longer timescale in Chapter 7. Perhaps one of the strands running through much of the book's material is a latent and at times extant utopianism, generally failed but necessary to the energy required for change in favour of recognition of human, social rather than economic needs, and glimpsed – as Benjamin argues (Chapter 3) – in ordinary things. The optimism of international modernism contains this, and it is present even in flagship cultural institutions claiming a universality of cultural value (Chapter 2). It grew from the diverse, migrant and in the end contained currents of modern art (Chapter 4) and equally from the explicitly urban setting for new cultural movements, from the new patterns of sociation enabled therein and from the emergence of bonds of common interest to replace those of kin or land (Chapter 1). This is all the summary of the book's arguments I will give, and in closing the text I think of Audre Lorde's argument that the master's tools will not dismantle the master's house (2000: 53–55). I have only the tools acquired in a somewhat variegated career, mainly imperfect, and an understanding (I hope) that the means do not point to but *produce* the ends. This is the case in writing as it is in social, cultural, economic and political formation and re-formation.

Notes for further reading

Chapters 7–9

The three chapters in this section share a worldwide focus on culture and society – I use the term *worldwide* in the sense of *mondialisation* (see Chapter 7) – in keeping with a move beyond the Euro-centric in the social sciences since the influence of post-colonial studies was added to that of feminism and Marxism in defining research agendas. Several recent books give insights into this framework as well as containing case studies from non-European sites: Jennifer Robinson's *Ordinary Cities* (2006 – cited in notes to Chapters 1 and 2) values the ordinary in any city; *Cosmopolitan Urbanism* (Binnie et al., 2006 – cited in notes to Chapters 3 and 4) examines living with difference in a range of contrasting situations; *Histories of the Future* (Rosenberg and Harding, 2005) investigates the effects of ubiquitous images and narratives of utopian, dystopian and technological futures on those who encounter them or are the subjects of such representations. Gary Bridge's *Reason in the City of Difference* (2005) approaches transactional rationality in context of theories and instances of difference. *Culture and Global Change* edited by Tracey Skelton and Tim Allen (1999) includes informative and critical essays on culture, globalisation, development, modernisation, local knowledges, ethnicity and resistance.

The literature of cultural intersections is not extensive, and I have drawn partly on my own experiences, but it includes at least two valuable sources: *Socialist Spaces* (Crowley and Reid, 2002); and *Architecture and Revolution* (Leach, 1999). I have also used *After the Fall*, edited by George Katsiaficas (2001); and cite Judd Stitziel's intriguing study of GDR fashion, *Fashioning Socialism* (2005). On the mass utopias of Soviet and capitalist cultural and socio-economic formations I recommend Susan Buck-Morss's excellent *Dreamworld and Catastrophe* (2002b), with insights on the comparative architectures of the two systems, and contemporary Russian art.

On the permeation of colonial and colonised cultural and social formation, Setha Low's *On the Plaza* (2000) gives an intricate critical reading of urban form in Spain and the Americas. Looking to everyday lives and adaptations of cultural forms in a globalised world, Fariba Adelkhah's *Being Modern in Iran* (1999) offers original insights, again negotiating the complexities of socio-cultural formations. *White Papers, Black Marks*, edited by Lesley Naa Norle Lokko (2000) includes a text on Buenos Aires, mainly concerned with the production of tango, but which I have cited in relation to interactions of the new and old worlds. The issue of cultural migrations and permeations among diasporas is not something I have covered, but is rising on cultural research agendas and dealt with in Nicholas Mirzoeff's *Diaspora and Visual Culture: Representing Africans and Jews* (2000).

On contemporary alternative, dissenting and activist art, Grant Kester's *Conversation Pieces* (2004) is the most coherently critical work in its field, and goes well beyond the contributions of Suzanne Lacy (1995) and Dolores Hayden (1995) – with a refreshingly European rather than North American focus. The catalogue to the 'Groundworks' exhibition (Kester, 2005) sets out a range of art projects for social and environmental justice.

I give sources for Lefebvre's writing in English translation in the notes to Chapters 1 and 2, and critical commentaries (Shields, 1999; Elden, 2004). In Chapter 9 I have used Lefebvre's *Writing on Cities* (1996), an anthology selected and translated by Eleonore Kofman and Elizabeth Lebas, which I find fluent and carefully introduced in an essay which is itself a contribution to the field. *Sarai Reader* is an annual compilation (available in print and on-line) which might exemplify the continuous cultural revolution Lefebvre glimpses in his urban theory, uneven in depth but diverse and questioning, mixing voices from India and the West (and elsewhere), generally with a bias to topical and cultural urban questions. Also in the terrain of contemporary, mainly net-based socio-cultural research and critique, *Economising Culture: On the (Digital) Culture Industry*, edited by Geoff Cox, Joasia Krysa and Anya Lewin (2004), includes essays on communism, cultures and commodity (Esther Leslie), self-organising communicative formations (Armin Medisch) and the spectacle (Adam Chmielewski). I end by recommending *Jeanne Hyvrard: Theorist of the Modern World* by Jennifer Waelti-Walters (1996), from which I quote in Chapter 7. As indicated, I find Hyvrard's writing difficult, and am sometimes lost in its fractured narratives or subdued by its insistent voices. But I found Lefebvre difficult at first (for different reasons), Adorno (obviously) and the poet Paul Celan, as well. From each I have derived some consolation – which, of course, is not the point.

Bibliography

Abu-Lughod, J. (1991) 'Going Beyond Global Babble' in King (1991), pp. 31–38

Adelkhah, F. (1999) *Being Modern in Iran*, London, Hurst & Co.

Adorno, T. W. (1991) *The Culture Industry: Selected Essays on Mass Culture*, London, Routledge

Adorno, T. W. (1994) *The Stars Down to Earth and Other Essays on the Irrational in Culture*, London, Routledge

Adorno, T. W. (1997) *Aesthetic Theory*, ed. G. Adorno and R. Tiedemann, trans. R. Hullot-Kentor, London, Athlone [first published in German, Frankfurt, Suhrkamp, 1970]

Adorno, T. W. and Horkheimer, M. (1997) *Dialectic of Enlightenment*, London, Verso [first published, 1947]

Adorno, T. W., Benjamin, W., Bloch, E., Brecht, B. and Lukács, G. (1980) *Aesthetics and Politics*, London, Verso

Ajuntament de Barcelona (1999) *Barcelona, Nous Projectes/Barcelona, New Projects*, Barcelona, Ajuntament de Barcelona [Catalan–English parallel text]

Ajuntament de Barcelona (2003a) *Building a New Sea Front for the New Millennium*, Barcelona, Infrastructures del Llevant de Barcelona SA

Ajuntament de Barcelona (2003b) *Besòs Seafront: A New Impetus for Barcelona*, Barcelona, Infrastructures del Llevant de Barcelona SA

Albrow, M. (1996) *The Global Age: State and Society Beyond Modernity*, Stanford (CA), Stanford University Press

Albrow, M. (1997) 'Travelling Beyond Local Cultures: Socioscapes in a Global City', in Eade (1997), pp. 37–55

Allen, J. (2002) 'Symbolic Economies: The Culturation of Economic Knowledge', in du Gay and Pryke (2002), pp. 39–58

Amin, A. (2002) 'Ethnicity and the Multicultural City, Living with Diversity', report for the Department of Transport, Local Government and the Regions, Durham, University of Durham

Anderson, N. (1923) *The Hobo*, Chicago, University of Chicago Press

Anderson, P. (1998) *The Origins of Postmodernity*, London, Verso

Angotti, T. (1993) *Metropolis 2000: Planning, Poverty and Politics*, London, Routledge

Appadurai, A. (1990) 'Disjuncture and Difference in the Global Cultural Economy', *Theory, Culture and Society*, vol. 7, 2/3, pp. 295–310

Appadurai, A., ed. (2001) *Globalization*, Durham (NC), Duke University Press

Araeen, R. (1991) 'From Primitivism to Ethnic Arts', in Hiller (1991), pp. 158–182

Arendt, H. (1958) *The Human Condition*, Chicago, University of Chicago Press

Arts Council of Great Britain (1989) *An Urban Renaissance*, London, Arts Council of Great Britain

Ashton, D. (1972) *The Life and Times of the New York School*, Bath, Adams and Dart

Augé, M. (1995) *Non-places: Introduction to an Anthropology of Supermodernity*, London, Verso

Azatyan, V. (2006) 'A Hole in the Sky', *The Internationaler*, June, pp. 8–9

Bailey, C., Miles, S. and Stark, P. (2004) 'Culture-led Urban Regeneration and the Revitalisation of Identities in Newcastle, Gateshead and the North East of England', *International Journal of Cultural Policy*, vol. 10, 10, pp. 47–65

Barber, S. (1995) *Fragments of the European City*, London, Reaktion

Barber, S. (2001) *Extreme Europe*, London, Reaktion

Barthes, R. (1982) *Empire of Signs*, trans. R. Howard, New York, Hill & Wang [first published in French, Geneva, Skira, 1970]

Baudelaire, C. (1958) [1868] *Flowers of Evil*, New York, New Directions

Baudelaire, C. (1975) *Selected Poems*, trans. and introduced Joanna Richardson, Harmondsworth, Penguin

Baudelaire, C. (1989) *Intimate Journals*, trans. Christopher Isherwood, London, Black Spring Press

Baudelaire, C. (1998) [1863] 'The Painter of Modern Life', in Harrison, Wood and Gaiger (1998), pp. 493–507

Bauman, Z. (1989) *Modernity and the Holocaust*, Cambridge, Polity

Bauman, Z. (1997) 'From Pilgrim to Tourist – or a Short History of Identity', in du Gay and Hall (1997), pp. 17–20

Bauman, Z. (1998) *Globalization: The Human Consequences*, Cambridge, Polity

Beall, J., ed. (1997) *A City for All: Valuing Difference and Working with Diversity*, London, Zed Books

Beckett, A. (1994) 'Take a Walk on the Wild Side', *Independent on Sunday*, 27 February, pp. 10–12

Beecher, J. and Bienvenu, R. (1983) *The Utopian Vision of Charles Fourier: Selected Works on Work, Love, and Passionate Attraction*, Columbia (MI), University of Missouri Press

Bell, D. and Haddour, A., eds (2000) *City Visions*, Harlow, Pearson Education

Bell, D. and Jayne, M., eds (2004) *City of Quarters – Urban Villages in the Contemporary City*, Aldershot, Ashgate

Bell, D. and Jayne, M., eds (2006) *Small Cities: Urban Experience beyond the Metropolis*, London, Routledge

Belsey, C. (1985) *The Subject of Tragedy: Identity and Difference in Renaissance Drama*, London, Routledge

Belsey, C. (2001) *Shakespeare and the Loss of Eden*, Basingstoke, Palgrave

Belsey, C. (2002) *Poststructuralism: A Very Short Introduction*, Oxford, Oxford University Press

Belsey, C. (2005) *Culture and the Real*, London, Routledge

Benjamin, A. ed. (1991) *The Problems of Modernity: Adorno and Benjamin*, London, Routledge

Benjamin, W. (1970) *Illuminations*, intro. Hannah Arendt, London, Fontana

Benjamin, W. (1997) *Charles Baudelaire*, trans. H. Zohm, London, Verso [first published in English, London, New Left Books, 1973; in German, as *Charles Baudelaire, Ein Lyriker im Zeitalter des Hochkapitalismus*, Frankfurt, Suhrkamp, 1969]

Benjamin, W. (1998) *Understanding Brecht*, London, Verso [first published as *Versuche über Brecht*, Frankfurt, Suhrkamp, 1966]

Benjamin, W. (1999) *The Arcades Project*, trans. H. Eiland and K. McLaughlin, Cambridge (MA), Harvard University Press [from *Das Passagen-Werk*, ed. R. Tiedman, Frankfurt, Suhrkamp, 1982]

Benjamin, W. (2003) *Understanding Brecht*, trans. A. Bostock, London, Verso [first published in German, Frankfurt, Suhrkamp, 1966; first published by Verso, 1998]

Bennett, J. (2001) *The Enchantment of Modern Life: Attachments, Crossings, and Ethics*, Princeton (NJ), Princeton University Press

Bennett S. and Butler, J., eds (2000) *Locality, Regeneration & Diversities*, Bristol, Intellect Books

Berg-Schlosser, D. and Kersting, N. (2003) *Poverty and Democracy: Self-help and Political Participation in Third World Cities*, London, Zed Books

La Berge, L. C. (2006) 'In for the Longue Durée: The Graduate Labour Struggle at NYU', *Radical Philosophy*, vol. 136, pp. 63–64

Bergen (2001) *Kulturby Bergen 2000, Norges Europeiske Kulturbyår: prosjektrapport, programdokumentasjon*, Bergen, Bergen City Council

Bergen (2004) 'The City of Bergen' [promotional leaflet], Bergen, City of Bergen Information Department

Berland, J. and Hornstein, S., eds (2000) *Capital Culture: A Reader on Modernist Legacies, State Institutions, and the Value(s) of Art*, Montréal, McGill-Queens University Press

Berman, M. (1983) *All That Is Solid Melts Into Air: The Experience of Modernity*, London, Verso

Betancour, A. and Hasdell, P. (2000) 'Tango: A Choreography of Urban Displacement', in Lokko (2000), pp. 146–175

Bethell, L., ed. (1987) *Colonial Spanish America*, Cambridge, Cambridge University Press

Bhargava, R. (2002) 'Responses to 9/11 – Individual and Collective Dimensions', *Sarai Reader*, 2, pp. 200–203

Bianchini, F. (1993) 'Culture and the Re-Making of European Cities', in Bianchini and Parkinson (1993), pp. 1–20

Bianchini, F. and Parkinson, M., eds (1993) *Cultural Policy and Urban Regeneration: The West European Experience*, Manchester, Manchester University Press

Bianchini, F., Fisher, M., Montgomery, J. and Worpole, K. (1988) *City Centres, City Cultures*, Manchester, Centre for Local Economic Studies

Bingaman, A., Sanders, L. and Zorach, R., eds (2002) *Embodied Utopias: Gender, Social Change and the Modern Metropolis*, London, Routledge

Binnie, J., Holloway, J., Milligan, S. and Young, C., eds (2006) *Cosmopolitan Urbanism*, London, Routledge

Bird, J. (1988) 'The Spectacle of Memory', in catalogue for Michael Sandle exhibition, London, Whitechapel Art Gallery, pp. 29–39

Bird, J. (1993) 'Dystopia on the Thames', in Bird *et al.* (1993), pp. 120–135

Bird, J., Curtis, B., Putnam, T., Robertson, G. and Tickner, L., eds (1993) *Mapping the Futures: Local Cultures, Global Change*, London, Routledge

Bloch, E. (1988) *The Utopian Function of Art and Literature*, trans. J. Zipes and F. Mecklenburg, Cambridge (MA), MIT Press

Bloch, E. (1991) *Heritage of Our Times*, trans. N. and S. Plaice, Cambridge (MA), MIT Press

Boal, A. (2000) *Theatre of the Oppressed*, London, Pluto [first published, 1979]

Booth, P. and Boyle, R. (1993) 'See Glasgow, See Culture', in Bianchini and Parkinson (1993), pp. 21–47

Borden, I. (2001) *Skateboarding, Space and the City: Architecture and the Body*, Oxford, Berg

Borden, I., Kerr, J., Rendell, J. with Pivaro, A., eds (2001) *The Unknown City: Contesting Architecture and Social Space*, Cambridge (MA), MIT Press

Bourdieu, P. (1984) *Distinction: A Social Critique of the Judgement of Taste*, trans. R. Nice, London, Routledge

Bourdieu, P. (1993) *The Field of Cultural Production*, ed. and trans. R. Johnson, Cambridge, Polity

Bourdieu, P. and Haacke, H. (1995) *Free Exchange*, Cambridge, Polity

Bové, J. and Dufour, F. (2001) *The World Is Not for Sale: Farmers Against Junk Food*, London, Verso

Boyle, M. and Hughes, G. (1991) 'The Politics of the Representation of the "Real" Discourses from the Left on Glasgow's Role as European City of Culture 1991', *Area*, vol. 23, pp. 217–228

Boyle, M. and Hughes, G. (1994) 'The Politics of Urban Entrepreneurialism in Glasgow', *Geoforum*, vol. 25, 4, pp. 453–470

Boyle, R. (1990) 'Regeneration in Glasgow: Stability, Collaboration and Inequity', in Judd and Parkinson (1990), pp. 109–132

Braunstein P. and Doyle, M. W., eds (2002) *Imagine Nation: The American Counterculture of the 1960s and '70s*, London, Routledge

Brettell, R. R. (1993) 'Camille Pissarro and Urban View Painting: An Introduction', in Stevens (1993), pp. xv–xxxvi

Bridge, G. (2005) *Reason in the City of Difference: Pragmatism, Communicative Action and Contemporary Urbanism*, London, Routledge

Broom, J. and Richardson, B. (1995) *The Self-Build Book*, 2nd edition, Totnes, Green Books

Brown, M. C. (1998) *Socialist Realist Painting*, London, Yale University Press

Buck-Morss, S. (1991) *The Dialectics of Seeing: Walter Benjamin and the Arcades Project*, Cambridge (MA), MIT Press

Buck-Morss, S. (2002a) 'A Global Public Sphere?' *Radical Philosophy*, vol. 111, pp. 2–10

Buck-Morss, S. (2002b) *Dreamworld and Catastrophe: The Passing of Mass Utopia on East and West*, Cambridge (MA), MIT Press

Bürger, P. (1984) *Theory of the Avant-Garde*, Minneapolis (MN), University of Minnesota Press

Burgess, E. W. (2003) 'The Growth of the City: An Introduction to a Research Project', in LeGates and Stout (2003), pp. 157–163 [first published in Park, Burgess and McKenzie (1967) [1925]]

Butler, J. (1993) *Bodies That Matter*, London, Routledge

Butler, J. (1997) *Excitable Speech: A Politics of the Performative*, London, Routledge

Byrne, D. (1997) 'Chaotic Places or Complex Places? Cities in a Post-Industrial Era', in Westwood and Williams (1997), pp. 50–70

Calnek, E. (1978) 'The Internal Structure of Cities in America: Pre-Columbian Cities: The Case of Tenochtitlan', in Schaedel, Hardoy and Kinzer (1978), pp. 315–326

Camacho, D. E. (1998) *Environmental Injustice, Political Struggles: Race, Class and the Environment*, Durham (NC), Duke University Press

Carmen, R. (1996) *Autonomous Development: Humanizing the Landscape*, London, Zed Books

Castells, M. (1996) *The Rise of the Network Society*, Oxford, Blackwell [2nd edition, 2000]

Cerdà, I. (1861) *Téoria de la viabilidad urbana y reforma de la Madrid*, unpublished report, Madrid [first published 1991, Madrid, National Institute of Administration, and Madrid City Council]

Chambers, I. (2001) *Culture after Humanism: History, Culture, Subjectivity*, London, Routledge

Chang, T. C. (2000) 'Renaissance Revisited: Singapore as a "Global City for the Arts"', *International Journal of Urban and Regional Research*, vol. 24, pp. 818–831

Childe, V. G. (2003) 'The Urban Revolution', in LeGates and Stout (2003), pp. 36–42 [first published in *Town Planning Review*, vol. 21, April 1950, pp. 3–17]

Chmielewski, A. J. (2004) 'The Spectacle: Global and Particular', in Cox, Krysa and Lewin (2004), pp. 225–249

Choay, F. (1997) *The Rule and the Model: On the Theory of Architecture and Urbanism*, Cambridge (MA), MIT Press

Cilliers, P. (1998) *Complexity & Postmodernism: Understanding Complex Systems*, London, Routledge

Clammer, J. (2003) 'Globalisation, Class, Consumption and Civil Society in South-east Asian Cities', *Urban Studies*, vol. 40, 2, pp. 403–419

Clark, T. J. (1973) *The Absolute Bourgeois: Artists and Politics in France 1848–1851*, London, Thames and Hudson

Cline, A. (1997) *A Hut of One's Own: Life Outside the Circle of Architecture*, Cambridge (MA), MIT Press

Coates, C. (2001) *Utopia Britannica: British Utopian Experiments 1325–1945*, London, Diggers & Dreamers Cooperative

Cochrane, G. (2000) 'Creating Tate Modern: 1996–2000', in Cole (2000), pp. 8–11

Cole, I., ed. (2000) *Beyond the Museum: Art, Institutions, People*, Oxford, Museum of Modern Art

Coles, A., ed. (1999) *The Optic of Walter Benjamin* [*de-, dis-, ex-*, vol. 3], London, Black Dog

Colomina, B., ed. (1992) *Sexuality & Space*, New York, Princeton Architectural Press

Colomina, B. (1996) *Privacy and Publicity: Modern Architecture and Mass Media*, Cambridge (MA), MIT Press

Connelly, M. (2000) *Void Moon*, New York, Orion

Connolly, W. (1991) *Identity/Difference*, Ithaca (NY), Cornell University Press

Coombes, A. (1991) 'Ethnography and the Formation of National and Cultural Identities', in Hiller (1991), pp. 189–214

Coombes, A, (1994a) *Reinventing Africa: Museums, Material Culture and Popular Imagination*, New Haven (CT), Yale University Press

Coombes, A. (1994b) 'Blinded by Science: Ethnography at the British Museum', in Pointon (1994), pp. 102–119

Cooper, D., ed. (1968) *The Dialectics of Liberation*, Harmondsworth, Penguin

Copjec, J. and Sorkin, M., eds (1999) *Giving Ground – The Politics of Propinquity*, London, Verso

Corbin, A. (1996) *The Foul & the Fragrant – Odour and the Social Imagination*, Basingstoke, Macmillan [first published in English, as *The Foul and the Fragrant: The Sense of Smell and its Social Image in Modern France*, Oxford, Berg, 1986]

Le Corbusier (1987) *The City of Tomorrow and Its Planning*, New York, Dover [from the 8th French edition, trans. F. Etchells, 1929, New York, Payson & Clarke, 1929]

Cortés, H. [1519–26], *Letters from Mexico*, trans. and ed. A. R. Pagden, New York, Grossman

Costello, D. (2000) 'The Work of Art and its Public: Heidegger and Tate Modern', in Cole (2000), pp. 12–26

Coutts-Smith, K. (1991) 'Some General Observations on the Problem of Cultural Colonialism', in Hiller (1991), pp. 14–31

Cox, G., Krysa, J. and Lewin, A. eds (2004) *Economising Culture: On the (Digital) Culture Industry*, New York, Autonomedia

Crawford, M. (1992) 'The World in a Shopping Mall', in Sorkin (1992), pp. 3–30

Cressey, P. G. (1932) *The Taxi-Hall Dancer*, Chicago, University of Chicago Press

Cresswell, T. (1996) *In Place Out of Place: Geography, Ideology, and Transgression*, Minneapolis (MN), University of Minnesota Press

Crouch, D. (2003) *The Art of Allotments: Culture and Cultivation*, Nottingham, Five Leaves

Crowley, D. and Reid, S. E., eds (2002) *Socialist Spaces: Sites of Everyday Life in the Eastern Bloc*, Oxford, Berg

Curtis, B. (2000) 'The Heart of the City', in Hughes and Sadler (2000), pp. 52–63

Curtis, K. (1999) *Our Sense of the Real: Aesthetic Experience and Arendtian Politics*, Durham (NC), Duke University Press

Daunton, M. and Hilton, M., eds (2001) *The Politics of Consumption: Material Culture and Citizenship in Europe and America*, Oxford, Berg

Davis, M. (1990) *City and Quartz*, London, Verso

Davis, M. (1998) *Ecology of Fear: Los Angeles and the Imagination of Disaster*, New York, Random House

Deamer, P. (1997) 'The Everyday and the Utopian', in Harris and Berke (1997), pp. 195–216

Dean, A. O. and Hursley, T. (2002) *Rural Studio: Samuel Mockbee and an Architecture of Decency*, New York, Princeton Architectural Press

Deepwell, K., ed. (1995) *New Feminist Art Criticism*, Manchester, Manchester University Press

Degen, M. (2002) 'Regenerating Public Life? A Sensory Analysis of Regenerated Public Places in el Raval, Barcelona', in Rugg and Hinchcliffe (2002), pp. 19–36

Degen, M. (2004) 'Barcelona's Games: the Olympics, Urban Design, and Global Tourism', in Sheller and Urry (2004), pp. 131–142

Delanty, G. (2003) *Community*, London, Routledge

Dethier, J. (1982) *Down to Earth, Adobe Architecture: An Old Idea, A New Future*, London, Thames and Hudson

Deutsche, R. (1991) 'Uneven Development: Public Art in New York City', in Ghirardo (1991), pp. 157–219 [first published, 1988, *October*, vol. 47, pp. 3–52]

Diaz del Castillo, B. (1963) *The Conquest of New Spain*, Harmondsworth, Penguin

Dicks, B. (2003) *Culture on Display: The Production of Contemporary Visitability*, Maidenhead, Open University Press

Didion, J. (2001) [1967] *Slouching Towards Bethlehem*, London, Fontana

Dodd, D. (1999) 'Barcelona, The Making of a Cultural City', in Dodd and van Hemel (1999), pp. 53–64

Dodd, D. and van Hemel, A., eds (1999) *Planning Cultural Tourism in Europe: A Presentation of Theories and Cases*, Amsterdam, Boekmanm Stichting

Donald, J. (1999) *Imagining the Modern City*, London, Athlone

Douglas, M. and Isherwood, B. (1979) *The World of Goods*, London, Allen Lane

Douglass, M. (1998) 'World City Formation in the Asia Pacific Rim: Poverty, Everyday Forms of Civil Society and Environmental Management', in Douglass and Friedmann (1998), pp. 107–137

Douglass, M. and Friedmann, J., eds (1998) *Cities for Citizens*, Chichester, Wiley

Dovey, K. (1999) *Framing Places: Mediating Power in Built Form*, London, Routledge

Dowling, R. (1993) 'Femininity, Place and Commodities: a Retail Case Study', *Antipode*, vol. 25, 4, pp. 295–319

Drew, B. (1998) *Crossing the Expendable Landscape*, St Paul (MN), Graywolf Press

Drewal, M. T. (2000) 'Nomadic Cultural Production in African Diaspora', in Mirzoeff (2000), pp. 115–142

Duncan, C. (1995) *Civilizing Rituals – Inside Public Art Museums*, London, Routledge

Durkheim, E. (1947) [1893] *The Division of Labour in Society*, Glencoe (IL), Free Press

Eade, J., ed. (1997) *Living the Global City: Globalization as Local Process*, London, Routledge

Eckardt, F. (2006) 'Urban Myth: The Symbolic Sizing of Weimar, Germany', in Bell and Jayne (2006), pp. 121–132

Edgell, S., Hetherington, K. and Warde, A., eds (1996) *Consumption Matters*, Oxford, Blackwell

Edwards, S., ed. (1973) *The Communards of Paris, 1871*, London, Thames and Hudson

Eisenschitz, A. (1997) 'The View from the Grassroots', in Pacione (1997), pp. 150–176

Elden, S. (2004) *Understanding Henri Lefebvre: Theory and the Possible*, London, Continuum

Engels, F. (1892) *The Condition of the Working Class in England in 1844*, London, Allen and Unwin

Etzioni, A. (2004) *From Empire to Community*, Basingstoke, Palgrave

Evans, G. (2001) *Cultural Planning: An Urban Renaissance?*, London, Routledge

Evans, G. (2003) 'Hard-branding the Cultural City: From Prado to Prada', *International Journal of Urban and Regional Research*, vol. 27, 2, pp. 417–440

Evans, G. (2004) 'Cultural Industry Quarters: From Pre-Industrial to Post-Industrial Production', in Bell and Jayne (2004), pp. 71–92

Evans, G. (2005) 'Measure for Measure: Evaluating the Evidence of Culture's Contribution to Regeneration', *Urban Studies*, vol. 42, 5/6, pp. 959–983

Extra]muros[(2002) *Capital do nada*, Lisbon, Extra]muros[and Callouste Gulbenkian Foundation

Fanon, F. (1967) *The Wretched of the Earth*, Harmondsworth, Penguin

Farrell, J. F. (1997) *The Spirit of the Sixties – The Making of Postwar Radicalism*, London, Routledge

Fathy, H. (1969) *Gourna – A Tale of Two Villages*, Cairo, Egyptian Ministry of Culture [republished as *Architecture for the Poor*, Chicago, University of Chicago Press, 1973, and Cairo, American University, 1989]

Featherstone, M. (1991) *Consumer Culture and Postmodernism*, London, Sage

Featherstone, M. (1995) *Undoing Culture: Globalization, Postmodernism and Identity*, London, Sage

Feaver, W. (1993) *Pitmen Painters: The Ashington Group 1934–1984*, Manchester, Carcanet

Feld, S. (2001) 'A Sweet Lullaby for World Music', in Appadurai (2001), pp. 189–216

Felshin, N., ed. (1995) *But Is It Art?*, Seattle (WA), Bay Press

Fennel, G. (1997) 'Local Lives – Distant Ties: Researching Community under Globalized Conditions', in Eade (1997), pp. 90–109

Fernandez, E. and Varley, A. (1998) *Illegal Cities: Law and Urban Change in Developing Countries*, London, Zed Books

Field, P. (1999) 'The Anti-Roads Movement: The Struggle of Memory Against Forgetting', in Jordan and Lent (1999), pp. 68–79

Finnegan, R. (1998) *Tales of the City: A Study of Narrative and Urban Life*, Cambridge, Cambridge University Press

Fischer, E. (1973) *Marx in His Own Words*, Harmondsworth, Penguin

Flyvbjerg, B. (1998) 'Empowering Civil Society: Habermas, Foucault and the Question of Conflict', in Douglass and Friedmann (1998), pp. 185–211

Foster, G. (1960) *Culture and Conquest: The American Spanish Heritage*, New York, Viking Fund Publications in Anthropology

Foster, H. (1985) *Recodings: Art, Spectacle, Cultural Politics*, Seattle (WA), Bay Press

Foster, H. (1996) *The Return of the Real: The Avant-Garde at the End of the Century*, Cambridge (MA), MIT Press

Foucault, M. (1967) *Madness and Civilization: A History of Insanity in the Age of Reason*, London, Tavistock [first published as *Histoire de la folie*, Paris, Libraire Plon, 1961]

Foucault, M. (1984) 'Le Retour de la morale', Interview, *Les Nouvelles*, 28 June [quoted in Dreyfus, H. and Rabinow, P. (1989) 'What Is Maturity', in Hoy, D. C., ed. (1989) *Foucault: A Critical Reader*, Oxford, Blackwell, pp. 109–21]

Foucault, M. (1988) *Politics, Philosophy, Culture*, ed. L. D. Kritzman, London, Routledge

Foucault, M. (1991) *Discipline and Punish: The Birth of the Prison*, Harmondsworth, Penguin

Foucault, M. (2000) 'The Subject and Power', in Nash (2000), pp. 8–26

Fowkes, R. (2002) 'The Role of Monumental Sculpture in the Construction of Socialist Spaces in Stalinist Hungary', in Crowley and Reid (2002), pp. 65–84

Frantz, D. and Collins, C. (1999) *Celebration USA – Living in Disney's Brave New Town*, New York, Henry Holt & Co.

Fraser, N. (1993) 'Rethinking the Public Sphere: A Contribution to the Critique of Actually Existing Democracy', in Robbins (1993), pp. 1–32

Fraser, N. (2001) 'Postcommunist Democratic Socialism?', in Katsiaficas (2001), pp. 200–202

Freire, P. (1972) *Pedagogy of the Oppressed*, Harmondsworth, Penguin

Fremion, Y. (2002) *Orgasms of History: 3000 Years of Spontaneous Insurrection*, Edinburgh, AK Press

Freud, S. (1930) *Civilization and its Discontents*, London, Hogarth Press

Friedberg, A. (1993) *Window Shopping*, Berkeley, University of California Press

Friedmann, J. (1998) 'The New Political Economy of Planning: The Rise of Civil Society', in Douglass and Friedmann (1998), pp. 19–35

Frisby, D. (1985) *Fragments of Modernity*, Cambridge, Polity

Frisby, D. (1992) *Simmel and Since: Essays on Georg Simmel's Social Theory*, London, Routledge

Frisby, D. and Featherstone, M., eds (1997) *Simmel on Culture*, London, Sage

Fuller, P. (1988) *Theorie – Art, and the Absence of Grace*, London, Chatto and Windus

Fyfe, N., ed. (1998) *Images of the Street: Planning, Identity and Control in Public Space*, London, Routledge

Le Galés, P. (1993) 'Rennes: Catholic Humanism and Urban Entrepreneurialism', in Bianchini and Parkinson (1993), pp. 178–198

Gandy, M. (2002) *Concrete and Clay: Reworking Nature in New York City*, Cambridge (MA), MIT Press

Garcia, B. (2002) 'City Fathers Win Medal for Style but Last Place in Social Events', *The Times Higher Education Supplement*, 11 October

Garcia, B. (2005) 'Deconstructing the City of Culture: The Long-term Cultural Legacies of Glasgow 1990', *Urban Studies*, vol. 42, 5/6, pp. 841–868

Gardiner, M. E. (2000) *Critiques of Everyday Life*, London, Routledge

Garnham, N. (1986) 'Bourdieu's *Distinction*', *Sociological Review*, 34 (May), pp. 423–433

Garreau, J. (1991) *Edge City: Life on the New Frontier*, New York, Doubleday

du Gay, P. and Hall, S., eds (1997) *Questions of Cultural Identity*, London, Sage du Gay, P. and Pryke, M., eds (2002) *Cultural Economy*, London, Sage

Generalitat de Catalunya (2001) *Cerdà: The Barcelona Extension (Eixample)*, Barcelona, Generalitat de Catalunya

George, S., Monbiot, G., German, L., Hayter, T., Callinicos, A. and Moody, K. (2001) *Anti-Capitalism: A Guide to the Movement*, London, Bookmarks

Ghannam, F. (1997) 'Re-imagining the Global: Relocation and Local Identities in Cairo', in Öncü and Weyland (1997), pp. 119–139

Ghirardo, D., ed. (1991) *Out of Site: A Social Criticism of Architecture*, Seattle (WA), Bay Press

Ghirardo, D. (1996) *Architecture After Modernism*, London, Thames and Hudson

Gilbert, A. (1998) 'World Cities and the Urban Future: The View From Latin America', in Lo and Yeung (1998), ch. 8

Gilloch, G. (1996) *Myth and Metropolis: Walter Benjamin and the City*, Cambridge, Polity

Gilloch, G. (2002) *Walter Benjamin: Critical Constellations*, Cambridge, Polity

Gilmore, A. (2004) 'Popular Music, Urban Regeneration and Cultural Quarters: The Case of the Rope Walks, Liverpool', in Bell and Jayne (2004), pp. 109–130

Gimeno, E. (2001) 'The Birth of the Barcelona Extension (*Eixample*)', in Generalitat de Catalunya (2001), pp. 20–23

Glancey, J. (2000) 'An Instructive Tale of Two Power Stations: His and Hers', *Guardian*, 17 May, p. 20

Goldfinger, M. (1993) *Villages in the Sun: Mediterranean Community Architecture*, New York, Rizzoli

Goldman, R. and Papson, S. (1998) *Nike Culture*, London, Thames and Hudson

González, A. and Lacuesta, R. (2002) *Barcelona: Architecture Guide 1929–2002* (revised edition), Barcelona, Gustavo Gili

Gonzalez, J. M. (1993) 'Bilbao: Culture, Citizenship and Quality of Life', in Bianchini and Parkinson (1993), pp. 73–89

Gottdiener, M. (2000) 'The Consumption of Space and the Spaces of Consumption', in Gottdiener, M. (ed.) *New Forms of Consumption: Consumers, Culture, and Commodification*, Lanham (MD), Rowman & Littlefield, pp. 265–286

Gough, J. and Eisenschitz, A. with McCulloch, A. (2006) *Spaces of Social Exclusion*, London, Routledge

Govil, N. (2002) 'The Metropolis and Mental Strife: The City in Science Fiction Cinema', *Sarai Reader*, 2, pp. 79–84

Gramsci, A. (1971) *Selections from the Prison Notebooks*, New York, International Publishers

Greenberg, C. (1986) *Perceptions and Judgments, 1939–1944*, vol. 1 of *Collected Essays and Criticism*, ed. John O'Brian, Chicago, University of Chicago Press

Greenhalgh, L. (1998) 'From Arts Policy to Creative Economy', *Media International Australia*, vol. 87 (May), pp. 84–94

Gregson, N. and Crewe, L. (2003) *Second-Hand Cultures*, Oxford, Berg

Griffin, R., ed. (1995) *Fascism*, Oxford, Oxford University Press

Grosz, E. (2004) *The Nick of Time – Politics, Evolution, and the Untimely*, Durham (NC), Duke University Press

Grunenberg, G. (1994) 'The Politics of Presentation: The Museum of Modern Art, New York', in Pointon (1994), pp. 192–211

Grūtas Park (2002) *Tiesa*, 1 April, Grūtas (Lithuania), Grūtas Park Publications

Guerrilla Girls (1995) *Confessions of the Guerrilla Girls*, New York, HarperCollins

Guha, R. (1989) *The Unquiet Woods: Ecological Change and Peasant Resistance in the Himalayas*, Oxford, Oxford University Press

Guha, R. and Martinez-Alier, J. (1997) *Varieties of Environmentalism: Essays North and South*, London, Earthscan

Gupta, S. (2002) *The Replication of Violence: Thoughts on International Terrorism after September, 2001*, London, Pluto

Habermas, J. (1981) *The Theory of Communicative Action*, 2 vols, trans. T. McArthy, Boston, Beacon Press [republished, Cambridge (MA), MIT Press, 1990]

Habermas, J (1987) *The Philosophical Discourse of Modernity*, Cambridge (MA), MIT Press

Habermas, J. (1989) *The Structural Transformation of the Public Sphere: An Inquiry into a Category of Bourgeois Society*, Cambridge (MA), MIT Press [first published in German, Darmstadt, Hermann Luchterhand Verlag, 1962]

Hall, P. (2001) *Cities of Tomorrow*, updated edition, Oxford, Blackwell [updated, 1996; first published, 1988]

Hall, P. and Ward, C. (1998) *Sociable Cities: The Legacy of Ebenezer Howard*, Chichester, Wiley

Hall, T. (1997) '(Re)placing the City: Cultural Relocation and the City as Centre', in Westwood and Williams (1997), pp. 202–218

Hall, T. (2004) 'Public Art, Civic Identity and the New Birmingham', in Kennedy (2004), pp. 63–71

Hall, T. and Hubbard, P., eds (1998) *The Entrepreneurial City: Geographies of Politics, Regime and Representation*, Chichester, Wiley

Hall, T. and Hubbard, P. (2001) 'Public Art and Urban Regeneration Advocacy, Claims and Critical Debates', *Landscape Research*, vol. 26, 1, pp. 5–26

Hall, T. and Smith, C. (2005) 'Public Art in the City: Meanings, Values, Attitudes and Roles', in Miles and Hall (2005), pp. 175–180

Hamburger, M. (1969) *The Truth of Poetry: Tensions in Modern Poetry from Baudelaire to the 1960s*, London, Weidenfeld & Nicolson

Hammersley, M. and Atkinson, P. (1995) *Ethnography: Principles in Practice*, 2nd edition, London, Routledge

Hannerz, U. (1980) *Exploring the City: Inquiries Towards an Urban Anthropology*, New York, Columbia University Press

Haque, S. (1988) 'The Politics of Space: The Experience of a Black Woman Architect', in Owusu (1988), pp. 55–60

Haraway, D. (1991) *Simians, Cyborgs and Women: The Reinvention of Nature*, London, Routledge

Harris, S. and Berke, D., eds (1997) *Architecture of the Everyday*, New York, Princeton Architectural Press

Harrison, C. and Wood, P., eds (1992) *Art in Theory, 1900–1992: An Anthology of Changing Ideas*, Oxford, Blackwell

Harrison, C., Wood, P. with Gaiger, J., eds (1998) *Art in Theory 1815–1900: An Anthology of Changing Ideas*, Oxford, Blackwell

Harutyunyan, A. (2004) 'Public Media Space' [exhibition leaflet], Yerevan (Armenia), Armenian Centre for Contemporary Experimental Art

Harvey, D. (1989) *The Urban Experience*, Baltimore (MD), Johns Hopkins University Press

Harvey, D. (2000) *Spaces of Hope*, Berkeley (CA), University of California Press

Harvey, P. (1996) *Hybrids of Modernity: Anthropology, the Nation State and the Universal Exhibition*, London, Routledge

Hayden, D. (1995) *The Power of Place: Urban Landscapes as Public History*, Cambridge (MA), MIT Press

ter Heide, H. and Wijnbelt, D. (1996) 'To Know and to Make: The Link Between Research and Urban Design', *Journal of Urban Design*, vol. 1, 1, pp. 75–90

Heidegger, M. (1959) *An Introduction to Metaphysics*, trans. Ralph Manheim, New Haven (CT), Yale University Press

Heidegger, M. (1984) *Early Greek Philosophy: The Dawn of Western Philosophy*, San Francisco, Harper & Row

Herder, J. G. von (1969) [1784–91] *Herder on Social and Political Culture*, ed. and trans. F. M. Barnard, Cambridge, Cambridge University Press

Hewitt + Jordan (2004) *I Fail to Agree*, Sheffield, Site Gallery

Hewitt + Jordan (2005) *The Neo-imperial Function* [project brochure], Sheffield, Hewitt + Jordan

Highmore, B. (2002a) *Everyday Life and Cultural Theory: An Introduction*, London, Routledge

Highmore, B., ed. (2002b) *The Everyday Life Reader*, London, Routledge

Hill, J., ed. (1998) *Occupying Architecture*, London, Routledge

Hiller, S., ed. (1991) *The Myth of Primitivism: Perspectives on Art*, London, Routledge

Hobbs, D. and May, T., eds (1993) *Interpreting the Field*, Oxford, Oxford University Press

Holding, N. (2003) *Armenia* [Bradt Travel Guide], Guilford (CT), Globe Pequot Press

hooks, b. (1995) *Art on My Mind: Visual Politics*, New York, The New Press

Hosagrahar, J. (2005) *Indigenous Modernities: Negotiating Architecture and Urbanism*, London, Routledge

House, J. (1979) 'The Legacy of Impressionism in France', in House and Stevens (1979), pp. 13–18

House, J. and Stevens, M. A., eds (1979) *Post-Impressionism: Cross-Currents in European Painting*, London, Royal Academy of Arts with Weidenfeld & Nicolson

Houtart, F. and Polet, F. (2001) *The Other Davos: The Globalization of Resistance to the World Economic System*, London, Zed Books

Howard, E. (1898) *Tomorrow! A Peaceful Path to Real Reform*, London, Swan Sonnenschein

Howard, E. (1946) *Garden Cities of Tomorrow*, London, Faber and Faber

Howard, E. (2003) 'The Town and Country Magnet', in LeGates and Stout (2003), pp. 313–316 [first published, 1898]

Hughes, J. and Sadler, S., eds (2000) *Non-Plan: Essays on Freedom, Participation and Change in Modern Architecture and Urbanism*, Oxford, Architectural Press

Hyvrard, J. (1975) *Les Prunes de Cythère*, Paris, Les Editions de Minuit

Hyvrard, J. (1989) *La Pensée corps*, Paris, Les Editions des Femmes

Illich, I. (1986) *H₂O and the Waters of Forgetfulness*, London, Marion Boyars

Ireland, C. (2004) *The Subaltern Appeal to Experience: Self-Identity, Late Modernity, and the Politics of Immediacy*, Montréal, McGill-Queens University Press

Irigaray, L. (1994) *Thinking the Difference: For a Peaceful Revolution*, London, Athlone

Irigaray, L. (1996) *I Love to You: Sketch of a Possible Felicity in History*, London, Athlone

Irigaray, L. (1999) *The Forgetting of Air*, London, Athlone

Isaak, J. A. (1996) *Feminism & Contemporary Art: The Revolutionary Power of Women's Laughter*, London, Routledge

Jackson, P. (1989) *Maps of Meaning*, London, Routledge

Jacobs, J. (1961) *The Death and Life of Great American Cities*, New York, Random House

Jacobs, J. (1969) *The Economy of Cities*, New York, Random House

Jacobus, J. (1973) *Matisse*, London, Thames and Hudson

Jameson, F. (1991) *Postmodernism, or the Cultural Logic of Late Capitalism*, London, Verso

Jamison, A. (2001) *The Making of Green Knowledge*, Cambridge, Cambridge University Press

Jayne, M. (2000) 'Imag(in)ing a Post-industrial Potteries', in Bell and Haddour (2000), pp. 12–26

Jessop, B. (1998) 'The Narrative of Enterprise and the Enterprise of Narrative: Place Marketing and the Entrepreneurial City', in Hall and Hubbard (1998), pp. 77–99

Johnson, H. (1999) 'Local Forms of Resistance: Weapons of the Weak', in Skelton and Allen (1999), pp. 159–166

Jones, J. (2000) 'Tate Modern Gets a Million Visitors in Just Six Weeks. Meanwhile at Poor Old Tate Britain . . .', *Guardian*, 28 June, arts section, p. 12

Jordan, T. and Lent, A. S., eds (2001) *Storming the Millennium: The New Politics of Change*, London, Lawrence & Wishart

Jowell, T. (2005) 'Why Should Government Support the Arts?', *Engage*, vol. 17, pp. 5–8 [address delivered 7 March 2005]

Jowett, R., Kain, J. P. and Baigeant, E. (1992) *The Cadastral Map in the Service of the State: A History of Property Mapping*, Chicago, University of Chicago Press

Judd, D. and Parkinson, M., eds (1990) *Leadership and Urban Regeneration*, Urban Affairs Annual Reviews, Newbury Park (CA), Sage

Jukes, P. (1990) *A Shout in the Street: The Modern Cities*, London, Faber and Faber

Kabeer, N. (1994) *Reversed Realities: Gender Hierarchies in Development Thought*, London, Verso

Kagarlitsky, G. (2001) 'The Road to Consumption', in Katsiaficas (2001), pp. 52–68

Kapferer, J. (2003) 'Culture and Commerce: European Culture Cities and Civic Distinction', in Miles and Kirkham (2003), pp. 31–42

Katsiaficas, G., ed. (2001) *After the Fall: 1989 and the Future of Freedom*, London, Routledge

Keating, M. (1988) *The City that Refused to Die – Glasgow: The Politics of Urban Representation*, Aberdeen, Aberdeen University Press

Keating, M. and Boyle, R. (1986) *Remaking Urban Scotland*, Edinburgh, Edinburgh University Press

Kelly, A. (2001) *Building Legible Cities*, Bristol, Bristol Legible City

Kennedy, L., ed. (2004) *Remaking Birmingham*, London, Routledge

Kerr, J. (2001) 'The Uncompleted Monument: London, War, and the Architecture of Remembrance', in Borden et al. (2001), pp. 68–89

Kester, G., ed. (1998) *Art, Activism, and Oppositionality: Essays from Afterimage*, Durham (NC), Duke University Press

Kester, G. (2004) *Conversation Pieces: Community and Communication in Modern Art*, Berkeley (CA), University of California Press

Kester, G., ed. (2005) *Groundworks: Environmental Collaboration in Contemporary Art* [exhibition catalogue], Pittsburgh (PA), Carnegie Mellon University

King, A. D., ed. (1991) *Culture, Globalization and the World System*, Basingstoke, Macmillan

King, A. D. (1994) *Global Cities: Post-Imperialism and the Internationalization of London*, London, Routledge

King, A. D., ed. (1996) *Re-Presenting the City: Ethnicity, Capital and Culture in the 21st-Century Metropolis*, Basingstoke, Macmillan

King, A. D. (2004) *Spaces of Global Cultures: Architecture Urbanism Identity*, London, Routledge

Klein, N. (2000) *No Logo*, London, Flamingo

Klein, N. N. (1997) *The History of Forgetting: Los Angeles and the Erasure of Memory*, London, Verso

Kohn, M. (2003) *Radical Space: Building the House of the People*, Ithaca (NY), Cornell University Press

Kotowski, B. and Frohling, M. (1993) *No Art, No City*, Berlin, Berufsverbandes Bildender Künstler

Kracauer, S. (1995) *The Mass Ornament: Weimar Essays*, trans. T. Y. Levin, Cambridge (MA), Harvard University Press

Kreivyte, L. (2005) 'Going Public: Strategies of Intervention in Lithuania', in Miles and Hall (2005), pp. 121–132

Kristeva, J. (2002) *Revolt, She Said*, Los Angeles, Semiotext(e)

Kroker, A. and Kroker, M., eds (1991) *Ideology and Power in the Age of Ruins*, New York, St Martin's Press

Kropotkin, P. (1915) [1902] *Mutual Aid, A Factor of Evolution*, London, Heinemann [popular edition revised from 1904 edition; first published in English, 1902]

Kumar, K. (1995) *From Post-industrial to Post-modern Society: New Theories of the Contemporary World*, Oxford, Blackwell

Kuspit, D. (1993) *The Myth of the Avant-Garde Artist*, Cambridge, Cambridge University Press

Laclau, E. (1996) *Emancipation(s)*, London, Verso

Laclau, E. and Mouffe, C. (1985) *Hegemony and Socialist Strategy: Towards a Radical Democratic Politics*, London, Verso

Lacour, C. B. (1996) *Lines of Thought: Discourse, Architectonics, and the Origin of Modern Philosophy*, Durham (NC), Duke University Press

Lacy, S., ed. (1995) *Mapping the Terrain*, Seattle (WA), Bay Press

Ladd, B. (1990) *Urban Planning and Civic Order in Germany, 1860–1914*, Cambridge, Cambridge University Press

Laforgue, J. (1956) *Selected Writings*, New York, Grove Press

Lai, C.-L. (2004) 'Art Exhibitions Travel the World', in Sheller and Urry (2004), pp. 90–102

Landry, C. (2000) *The Creative City: A Toolkit for Urban Innovators*, London, Earthscan

Landry, C. and Bianchini, F. (1995) *The Creative City*, London, Demos

Laporte, D. (2000) *History of Shit*, Cambridge (MA), MIT Press [first published as *Histoire de la merde (Prologue)*, Paris, Christian Bourgeois, 1993]

Larson, G. O. (1997) *American Canvas*, Washington DC, National Endowment for the Arts

Lash, S. and Urry, J. (1994) *Economies of Signs and Space*, London, Sage

Latour, B. (2004) *Politics of Nature: How to Bring the Sciences into Democracy*, Cambridge (MA), Harvard University Press

Leach, N., ed. (1999) *Architecture and Revolution: Contemporary Perspectives on Central and Eastern Europe*, London, Routledge

Leadbeater, B. J. R. and Way, N., eds (1996) *Urban Girls: Resisting Stereotypes, Creating Identities*, New York, New York University Press

Leavitt, J. (1994) 'Planning in an Age of Rebellion: Guidelines to Activist Research and Applied Planning', *Planning Theory*, vol. 10/11, pp. 111–130

Lee, M. J. (1993) *Consumer Culture Reborn: The Cultural Politics of Consumption*, London, Routledge

Lefebvre, H. (1970) *La Révolution urbaine*, Paris, Gallimard

Lefebvre, H. (1989) *Le Somme et le reste*, Paris, Méridiens Klincksieck [first published, Paris, La Nef de Paris, 1959]

Lefebvre, H. (1991) *The Production of Space*, Oxford, Blackwell

Lefebvre, H. (1992) *Critique of Everyday Life* vol. 1, London, Verso

Lefebvre, H. (1995) *Introduction to Modernity*, trans. J. Moore, London, Verso

Lefebvre, H. (1996) *Writings on Cities*, trans. E. Kofman and E. Lebas, Oxford, Blackwell

Lefebvre, H. (2000) *Everyday Life in the Modern World*, London, Athlone

Lefebvre, H. (2003) *The Urban Revolution*, foreword N. Smith, London, Routledge

LeGates, R. T. and Stout, F., eds (2003) *The City Reader*, 3rd edition, London, Routledge

Leslie, E. (2000) *Walter Benjamin: Overpowering Conformism*, London, Pluto

Leslie, E. (2001) 'Tate Modern: A Year of Sweet Success', *Radical Philosophy*, vol. 109. pp. 2–5

Lever, W. and Moore, C., eds (1986) *The City in Transition: Public Policies and Agencies for the Economic Regeneration of Clydeside*, Oxford, Clarendon Press

Levinson, S. (1998) *Written in Stone: Public Monuments in Changing Societies*, Durham (NC), Duke University Press

Lewandowska, M. and Cummings, N. (2004) *Enthusiasts from Amateur Film Clubs*, Warsaw, Centre for Contemporary Art

Lewis, M. (1991) 'What Is to Be Done?' in Kroker and Kroker (1991), pp. 3–5

Ley, D. (1996) *The New Middle Class and the Remaking of Central City*, Oxford, Oxford University Press

Ley, D. and Olds, K. (1988) 'Landscape as Spectacle: World Fairs and the Culture of Heroic Consumption', *Environment and Planning D: Society and Space*, vol. 6, pp. 191–212

Liebeschuetz, W. (1992) 'The End of the Ancient City', in Rich (1992), pp. 1–49

Light, A. and Smith, J. M., eds (2005) *The Aesthetics of Everyday Life*, New York, Columbia University Press

Lim, H. (1993) 'Cultural Strategies for Revitalizing the City: A Review and Evaluation', *Regional Studies*, vol. 27, 6, pp. 589–594

Lippard, L. (1997) *The Lure of the Local*, New York, The New Press

Lloyd, R. (2006) *Neo-Bohemia: Art and Commerce in the Post-Industrial City*, London, Routledge

Lo, F. and Yeung, Y., eds (1998) *Globalisation and the World of Large Cities*, Tokyo, UN University Press

Lodziak, C. (2002) *The Myth of Consumerism*, London, Pluto

Loftman, P. and Nevin, B. (1998) 'Pro-growth Local Economic Development Strategies: Civic Promotion and Local Needs in Britain's Second City', in Hall and Hubbard (1998), pp. 129–148

Loftman, P. and Nevin, B. (2003) 'Prestige Projects, City-Centre Restructuring and Social Exclusion: Taking the Long-Term View', in Miles and Hall (2003), pp. 76–91

Lokko, L. N. N., ed. (2000) *White Papers, Black Marks: Architecture, Race, Culture*, London, Athlone

Lorde, A. (2000) 'The Master's Tools Will Never Dismantle the Master's House', in Rendell, Penner and Borden (2000), pp. 53–55 [address to the Second Sex Conference, New York, 29 September 1979]

Low, S. (2000) *On the Plaza: The Politics of Public Space and Culture*, Austin (TX), University of Texas Press

Lury, C. (1996) *Consumer Culture*, Cambridge, Polity [first published, New Brunswick (NJ), Rutgers University Press, 1992]

Lynch, K. (1981) *Good City Form*, Cambridge (MA), MIT Press

MacAvera, B. (1990) *Art, Politics and Ireland*, Dublin, Open Aire

MacCannell, D. (1976) *The Tourist: A New Theory of the Leisure Class*, New York, Schocken Books

MacCannell, D. (1996) *Tourist or Traveller?*, London, BBC Education

MacCannell, D. (1999) 'New Urbanism and its Discontents', in Copjec and Sorkin (1999), pp. 106–130

MacDonald, D. (1953) 'Profiles: Action on West Fifty-Third Street – I', *The New Yorker*, 12 December, p. 49

McElligot, A. (2001) *The German Urban Experience 1900–1945*, London, Routledge

McEwen, I. K. (1993) *Socrates' Ancestor: An Essay on Architectural Beginnings*, Cambridge (MA), MIT Press

McKay, G. (1996) *Senseless Acts of Beauty: Cultures of Resistance since the Sixties*, London, Verso

McLay, F. (1990) *The Reckoning*, Glasgow, Workers City

McQuire, S. (2005) 'The Burden of Culture in the Global City', in McQuire and Papastergiadis (2005), pp. 204–209

McQuire, S. and Papastergiadis, N., eds (2005) *Empires, Ruins and Networks: The Transcultural Agenda in Art*, London, Rovers Oram Press

Madsen, P. and Plunz, R., eds (2002) *The Urban Lifeworld: Formation, Perception, Representation*, London, Routledge

Maisels, C. K. (1993a) *The Emergence of Civilization: From Hunting and Gathering to Agriculture, Cities, and the State in the Near East*, London, Routledge

Maisels, C. K. (1993b) *The Near East: Archaeology in the 'Cradle of Civilization'*, London, Routledge

Malkin, I. (1987) *Religion and Colonization in Ancient Greece*, Leiden, Brill

Marcus, L. and Nead, L., eds (1998) *The Actuality of Walter Benjamin*, London, Lawrence & Wishart

Marcuse, H. (1968) 'Liberation from the Affluent Society', in Cooper(1968), pp. 175–192

Marcuse, H. (1969) *An Essay on Liberation*, Harmondsworth, Penguin

Marcuse, H. (1978) *The Aesthetic Dimension*, Boston, Beacon Press

Marcuse, H. (1998) *Technology, War and Fascism* (*Collected Papers*, vol. 1), ed. D. Kellner, London, Routledge

Marcuse, H. (2001) *Towards a Critical Theory of Society* (*Collected Papers*, vol. 2), ed. and intro. D Kellner, London, Routledge

Marcuse, H. (2005) *The New Left and the 1960s* (*Collected Papers*, vol. 3), ed. and intro. D. Kellner, London, Routledge

Marcuse, P. (2002) 'The Layered City', in Madsen and Plunz (2002), pp. 94–114

Marris, P. (1998) 'Planning and Civil Society in the Twenty-first Century: An Introduction', in Douglass and Friedman (1998), pp. 9–18

Martin, C. J. (1999) 'The New Migrants: Flexible Workers in a Global Economy', in Skelton and Allen (1999), pp. 180–189

Marvila (nd) 'Marvila' [leaflet and map, parallel Portuguese and English texts], Marvila, Junta de Freguesia

Marx, K. (1947) *The German Ideology*, ed. R. Pascal, New York, International Publishers

Massey, D. (1994) *Space, Place and Gender*, Cambridge, Polity

Massey, D. (2005) *For Space*, London, Sage

Maxwell, K. (2002) 'Lisbon: The Earthquake of 1755 and Urban Recovery under the Marquês de Pombal', in Ockman (2002), pp. 20–45

Medosch, A. (2004) 'Society in Ad-hoc Mode: Decentralised, Self-organising, Mobile', in Cox, Krysa and Lewin (2004), pp. 135–161

Mehta, R. (1997) 'Mediation and Participation in a Delhi Slum', in Beall (1997), pp. 261–267

Mellaart, J. (1967) *Çatal Hüyük*, London, Thames and Hudson

Meller, H. E., ed. (1979) *The Ideal City*, Leicester, Leicester University Press

Meller, H. E. (2001) *European Cities 1890s–1930s: History, Culture and the Built Environment*, Chichester, Wiley

Merrifield, A. and Swyngedouw, E., eds (1996) *The Urbanization of Injustice*, London, Lawrence & Wishart

Meskimmon, M. (1997) *Engendering the City: Women Artists and Urban Spaces*, London, Scarlet Press

Michalski, S. (1998) *Public Monuments: Art in Political Bondage*, London, Reaktion

Michelson, A. (1976) 'Reading Einstein Reading Capital', *October*, vol. 2 (Summer), pp. 27–38 [part 2, *October*, vol. 3 (Spring 1977), pp. 82–88]

Miles, M. (1997) *Art, Space & the City*, London, Routledge

Miles, M. (1998) 'A Game of Appearance: Public Art and Urban Development – Complicity or Sustainability?', in Hall and Hubbard (1998), pp. 203–224

Miles, M. (2000) *The Uses of Decoration: Essay in the Architectural Everyday*, Chichester, Wiley

Miles, M. (2004) *Urban Avant-Gardes: Art, Architecture and Change*, London, Routledge

Miles, M. (2005) 'Interruptions: Testing the Rhetoric of Culturally Led Urban Development', *Urban Studies* vol. 42, 5/6 (May), pp. 889–911

Miles, M. (2006) 'Utopias of Mud: Hassan Fathy and Alternative Modernisms', *Space & Culture*, vol. 9 , 2, pp. 116–139

Miles, M. and Hall, T., eds (2003) *Urban Futures: Critical Commentaries on Shaping the City*, London, Routledge

Miles, M. and Hall, T., eds (2005) *Interventions*, Bristol, Intellect Books

Miles, M. and Kirkham, N., eds (2003) *Cultures & Settlements*, Bristol, Intellect Books

Miles, M. and Miles, S. (2004) *Consuming Cities*, Basingstoke, Palgrave

Miles, M., Hall, T., and Borden, I., eds (2003) *The City Cultures Reader*, 2nd revised edition, London, Routledge

Miles, S. (1998) *Consumerism as a Way of Life*, London, Sage

Miles, S. (2003) 'Resistance or Security? Young People and the Appropriation of Urban, Cultural and Consumer Space', in Miles and Hall (2003), pp. 65–75

Miller, D. (1977) 'Idelfons Cerdà, an Introduction', *Architectural Association Quarterly*, vol. 9, p. 1

Miller, D. (1998) *A Theory of Shopping*, Ithaca (NY), Cornell University Press

Miller, D. (2001) *The Dialectics of Shopping*, Chicago, University of Chicago Press

Minh-ha, T. T. (1994) 'Other than Myself/My Other Self', in Robertson et al. (1994), pp. 9–28

Mirzoeff, N., ed. (2000) *Diaspora and Visual Culture: Representing Africans and Jews*, London, Routledge

Mishan, E. (1969) *The Costs of Economic Growth*, Harmondsworth, Penguin

Mockbee, S. (1998) 'The Rural Studio', *Architectural Design*, vol. 68, 7/8 (July–August), pp. 72–79

Molotch, H. (1976) 'The City as a Growth Machine', *American Journal of Sociology*, vol. 82, pp. 309–332

Molyneux, J. (2001) *Rembrandt & Revolution*, London, Redwords

Mooney, G. (2004) 'Cultural Policy as Urban Transformation? Critical Reflections on Glasgow, European City of Culture, 1990', *Local Economy*, vol. 19, 4, pp. 327–340

Morse, R. (1987) 'Urban Development', in Bethell (1987), pp. 165–202

Mouffe, C. (2000) *The Democratic Paradox*, London, Verso

Mozingo, L. (1989) 'Women and Downtown Open Spaces', *Place*, vol 6, 1, pp. 38–47

Mulhern, F. (2000) *Culture/Metaculture*, London, Routledge

Mulvey, L. (1999) 'Reflections on Disgraced Monuments', in Leach (1999), pp. 228–233

Mumford, L. (1937) 'What Is a City?', *Architectural Record*, LXXXII (November), in LeGates and Stout (2003), pp. 93–96

Mumford, L. (1946) 'The Garden City Idea and Modern Planning', in Howard (1946), pp. 29–40

Mumford, L. (1955) *The Human Prospect*, London, Secker and Warburg

Mumford, L. (1961) *The City in History*, New York, Harcourt Brace Jovanovich

Muttit, G. and Marriott, J. (2004) *Some Common Concerns: Imagining BP's Azerbaijan-Georgia-Turkey Pipelines System*, London, PLATFORM with Cornerhouse, Friends of the Earth, Campagna per la Riforma della Banca Mondiale, CEE Bankwatch Network and the Kurdish Human Rights Project

Myerscough, J. (1988) *The Economic Importance of the Arts in Britain*, London, Policy Studies Institute

Myerscough, J. (1994) *European Cities of Culture and Cultural Months*, Glasgow, The Network of Cultural Cities of Europe

Myrvoll, S. (1999) 'Cultural Heritage Tourism in Norway, with the Focus on Bergen', in Dodd and van Hemel (1999), pp. 44–52

Naegele, K. (1958) 'Attachment and Alienation: Contemporary Aspects of the Work of Durkheim and Simmel', *American Journal of Sociology*, vol. 63

Nash, K., ed. (2000) *Readings in Contemporary Political Sociology*, Oxford, Blackwell

Nava, M. (1992) *Changing Cultures: Feminism, Youth and Consumerism*, London, Sage

Nava, M. and O'Shea, A., eds (1996) *Modern Times: Reflections on a Century of English Modernity*, London, Routledge

Notes from Nowhere (2003) *We Are Everywhere: The Irresistible Rise of Global Anticapitalism*, London, Verso

O'Connor, J. (1998) 'Popular Culture, Cultural Intermediaries and Urban Regeneration', in Hall and Hubbard (1998), pp. 225–240

O'Connor, J. and Wynne, D., eds (1996) *From the Margins to the Centre: Cultural Production and Consumption in the Post-Industrial City*, Aldershot, Ashgate

O'Shea, A. (1996) 'English Subjects of Modernity', in Nava and O'Shea (1996), pp. 7–37

Ockman, J., ed. (2002) *Out of Ground Zero: Case Studies in Urban Reinvention*, Munich, Prestel Verlag

Oh, M. and Arditi, J. (2000) 'Shopping and Postmodernism: Consumption, Production, Identity, and the Internet', in Gottdiener (2000), pp. 71–89

Olds, K. and Yeung, Y. (2004) 'Pathways to Global City Formation: Views from the Developmental State of Singapore', *Review of International Political Economy*, vol. 11, pp. 489–521

Öncü, A. and Weyland, P., eds (1997) *Space, Culture and Power: New Identities in Globalizing Cities*, London, Zed Books

Owusu, K., ed. (1988) *Storms of the Heart: An Anthology of Black Arts & Culture*, London, Camden Press

Pacione, P., ed. (1997) *Britain's Cities: Geographies of Division in Urban Britain*, London, Routledge

Pahl, R. E. (1975) *Whose City? and Further Essays on Urban Society*, Harmondsworth, Penguin

Papanek, V. (1984) *Design for the Real World: Human Ecology and Social Change*, London, Thames and Hudson

Papanek, V. (1995) *The Green Imperative: Ecology and Ethics in Design and Architecture*, London, Thames and Hudson

Parekh, B. (2000) *Rethinking Multiculturaslism*, Basingstoke, Macmillan

Park, R. (1967) 'The City: Suggestions for the Investigation of Human Behaviour in an Urban environment', in Park, Burgess and McKenzie (1967) [first published, 1916, *American Journal of Sociology*, vol. 20, pp. 577–612]

Park, R., Burgess, E. W. and McKenzie, R. (1967) [1925] *The City*, intro. Morris Janowitz, Chicago, University of Chicago Press

Paz, O. (1985) *The Labyrinth of Solitude*, Harmondsworth, Penguin [first published as *El Labertino de la Soledad*, Mexico City, Fondo de Culturo Económica, 1959]

Pearson, N. (1982) *The State and the Visual Arts*, Milton Keynes, Open University Press

Peet, R. and Watts, M. (1996) *Liberation Ecologies: Environment, Development, Social Movements*, London, Routledge

Pessoa, F. (2001) *The Book of Disquiet*, trans. R. Zenith, London, Penguin

Phillips, P. (1988) 'Out of Order: The Public Art Machine', *Artforum* (December), pp. 92–96

Phillips, P. (1995) 'The Private Is Public: Peggy Diggs and the System', in Felshin (1995), pp. 283–308

Pickles, J. (2004) *A History of Spaces: Cartographic Reason, Mapping and the Geo-coded World*, London, Routledge

Pile, S., Brook, C. and Mooney, G., eds (1999) *Unruly Cities?*, London, Routledge, and Buckingham, Open University Press

Pinder, D. (2005) *Visions of the City*, Edinburgh, Edinburgh University Press

Plows, A. (1995a) 'The Donga Tribe: Practical Paganism Comes Full Circle', *Creative Mind*, vol. 27 (Summer), p. 26

Plows, A. (1995b) 'Ecophilosophy and Popular Protest: The Significance and Implications of the Ideology and Actions of the Donga Tribe', *Alternative Futures and Popular Protest*, vol. 1, np

Pointon, M., ed. (1994) *Art Apart: Art Institutions and Ideology Across England and North America*, Manchester, Manchester University Press

Pollock, G. (1988) *Vision and Difference: Femininity, Feminism, and Histories of Art*, London, Routledge

Raban, J. (1974) *Soft City*, London, Fontana

Rabinow, P. (1989) *French Modern: Norms and Forms of Social Environment*, Cambridge (MA), MIT Press

Rama, A. (1996) *The Lettered City*, Durham (NC), Duke University Press

Rancière, J. (2004) *The Politics of Aesthetics*, London, Continuum

Raven, A. (1993) *Art in the Public Interest*, New York, Da Capo Press

Ray, M.-A. (1997) '*Gecekondu*', in Harris and Berke (1997), pp. 153–165

Rendell, J. (1998) 'Doing It (Un)Doing It. (Over)Doing It Yourself; Rhetorics of Architectural Abuse', in Hill (1998), pp. 229–246

Rendell, J. (1999) 'Thresholds, Passages and Surfaces: Touching, Passing and Seeing in the Burlington Arcade', in Coles (1999), pp. 168–191

Rendell, J. (2002) *The Pursuit of Pleasure: Gender, Space and Architecture in Regency London*, London, Athlone

Rendell, J., Penner, B. and Borden, I., eds (2000) *Gender Space Architecture: An Interdisciplinary Introduction*, London, Routledge

Retort [Iain Boal, T. J. Clark, Joseph Mathews, Michael Watts] (2005) *Afflicted Powers: Capital and the Spectacle in the Age of War*, London, Verso

Rich, J., ed. (1992) *The City in Late Antiquity*, London, Routledge

Richards, G. (1991) *The Philosophy of Gandhi*, Richmond, Curzon Press

Richards, G. (1999) 'European Cultural Tourism: Patterns and Prospects', in Dodd and van Hemel (1999), pp. 16–32

Richards, J. M., Serageldin, I. and Rastorfer, D. (1985) *Hassan Fathy*, Singapore, Concept Media, and London, Architectural Press

Ridgeway, J. (2004) *It's All for Sale: The Control of Global Resources*, Durham (NC), Duke University Press

Ritzer, G. (1999) *Consumption and the Re-enchantment of Everyday Life*, Thousand Oaks (CA), Pine Forge Press

Robbins, B., ed. (1993) *The Phantom Public Sphere*, Minneapolis (MN), University of Minnesota Press

Robbins, E. (1996) 'Thinking Space/Seeing Space: Thamesmead Revisited', *Urban Design International*, vol. 1, 3, pp. 283–291

Roberts, J. (2001) 'Art, Politics and Provincialism', *Radical Philosophy*, vol. 106, pp. 2–6

Robertson, G., Mash, M., Tickner, L., Bird, J., Curtis, B. and Putnam, T., eds (1994) *Travellers' Tales: Narratives of Home and Displacement*, London, Routledge

Robertson, R. (1992) *Globalization*, London, Sage

Robinson, H., ed. (2001) *Feminism-Art-Theory: An Anthology 1968–2000*, Oxford, Blackwell

Robinson, J. (2006) *Ordinary Cities: Between Modernity and Development*, London, Routledge

Rodgers, P. (1989) *The Work of Art* [summary of John Myerscough, *The Economic Importance of the Arts in Britain*, 1988], London, Policy Studies Institute

Rogoff, I. (2000) *Terra Infirma: Geography's Visual Culture*, London, Routledge

Roschelle, A. R. and Wright, T. (2003) 'Gentrification and Social Exclusion: Spatial Policing and Homeless Activist Responses in the San Francisco Bay Area', in Miles and Hall (2003), pp. 149–166

Rosenberg, D. and Harding, S., eds (2005) *Histories of the Future*, Durham (NC), Duke University Press

Roy, A. (2001) *Power Politics*, Cambridge (MA), South End Press

Rugg, J. (2002) 'Budapest's Statue Park: Collective Memory or Collective Amnesia?', unpublished conference paper

Rugg, J. and Hinchcliffe, D., eds (2002) *Recoveries and Reclamations*, Bristol, Intellect Books

Rugg, J. and Sedgwick, M. (2001) 'Budapest's Statue Park: Memorial or Counter-monument?' *Soundings*, vol. 17, pp. 94–112

Ryan, J. (1994) 'Women, Modernity and the City', *Theory, Culture and Society*, vol. 11, 4 (November), pp. 35–64

Sadler, S. (1999) *The Situationist City*, Cambridge (MA), MIT Press

Salecl, R. (1999) 'The State as a Work of Art: The Trauma of Ceauçescu's Disneyland', in Leach (1999), pp. 92–111

Sandercock, L. (1998) *Towards Cosmopolis*, Chichester, Wiley

Sandercock, L. (2006) 'Cosmopolitan Urbanism: A Love Song to our Mongrel Cities', in Binnie et al. (2006), pp. 37–52

Santos, D. (2002) 'Da singularidade na condiçao pós-moderna, ou e mediaçao possivel da arte contemporânea', in Extra]muros[(2002), pp. 160–169 [parallel Portuguese–English text]

Sassen, S. (1991) *The Global City: New York, London, Tokyo*, Princeton (NJ), Princeton University Press

Savage, M. and Warde, A. (1993) *Urban Sociology, Capitalism and Modernity*, Basingstoke, Macmillan

Schaedel, R. P., Hardoy, J. E. and Kinzer, N. S., eds (1978) *Urbanization in the Americas from its Beginnings to the Present*, Den Haag, Mouton

Schelling, V. (1999) 'The People's Radio of Vila Nossa Senhora Aparecida: Alternative Communication and Cultures of Resistance in Brazil', in Skelton and Allen (1999), pp. 167–179

Schivelbusch, W. (1998) *In a Cold Crater: Cultural and Intellectual Life in Berlin, 1945–1948*, Berkeley (CA), University of California Press

Schor, N. (1992) 'Cartes Postales: Representing Paris 1900', *Critical Inquiry*, vol. 18, pp. 188–241

Sciorra, J. (1996) 'Return to the Future: Puerto Rican Vernacular Architecture in New York City', in King (1996), pp. 60–92

Scott, A. J. (2000) *The Cultural Economy of Cities*, London, Sage

Seabrook, J. (1993) *Pioneers of Change: Experiments in Creating a Humane Society*, Gabriola Island (BC), New Society Publishers

Seabrook, J. (1996) *In the Cities of the South: Scenes from a Developing World*, London, Verso

Seiler, C. (2000) 'The Commodification of Rebellion: Rock Culture and Consumer Capitalism', in Gottdiener (2000), pp. 203–226

Selwood, S. (1995) *The Benefits of Public Art*, London, Policy Studies Institute

Sennett, R. (1970) *The Uses of Disorder*, New York, Norton

Sennett, R. (1990) *The Conscience of the Eye*, New York, Norton

Sennett, R. (1992) [1974] *The Fall of Public Man*, New York, Norton

Sennett, R. (1995) *Flesh and Stone: The Body and the City in Western Civilization*, London, Faber and Faber

Sennett, R. (1998) *The Corrosion of Character: The Personal Consequences of Work in the New Capitalism*, New York, Norton

Sezneva, O. (2002) 'Living in the Russian Present with a German Past: The Problems of Identity in the City of Kaliningrad', in Crowley and Reid (2002), pp. 47–64

Shane, G. (2002) 'The Machine in the City: Phenomenology and Everyday Life in New York', in Madsen and Plunz (2002), pp. 218–236

Sharp, J., Pollock, V. and Paddison, R. (2005) 'Just Art for a Just City: Public Art and Social Inclusion in Urban Regeneration', *Urban Studies*, vol. 42, 5/6, pp. 1001–1023

Shattuck, R. (1969) *The Banquet Years: The Origins of the Avant-Garde in France: 1885 to World War I*, London, Jonathan Cape

Sheller, M. and Urry, J., eds (2004) *Tourism Mobilities: Places to Play, Places in Play*, London, Routledge

Sheridan, S., ed. (1988) *Grafts: Feminist Cultural Criticism*, London, Verso

Shields, R. (1992) *Lifestyle Shopping: The Subject of Consumption*, London, Routledge

Shields, R. (1999) *Lefebvre, Love & Struggle*, London, Routledge

Short, J. R. (1996) *The Urban Order: An Introduction to Cities, Culture, and Power*, Oxford, Blackwell

Sibley, D. (1995) *Geographies of Exclusion*, London, Routledge

Sibley, D. (2001) 'The Binary City', *Urban Studies*, vol. 38, 2, pp. 239–250

Sider, G. M. (1989) *Culture and Class in Anthropology and History*, Cambridge, Cambridge University Press

Siegert, H. and Stern, R. (2002) 'Berlin: Film and the Representation of Urban Reconstruction since the Fall of the Wall', in Ockman (2002), pp. 116–137

Sieverts, T. (2003) *Cities Without Cities: An Interpretation of the Zwischenstadt*, London, Spon

Simmel, G. (1969) *The Sociology of Georg Simmel*, ed. K. H. Wolff, New York, Free Press [first published, 1950]

Simmel, G. (1990) *The Philosophy of Money*, ed. D. Frisby, 2nd edition, London, Routledge [originally published as *Philosophie des Geldes*, 2nd edition, Berlin, 1907]

Simmel, G. (1997) *Simmel on Culture*, ed. D. Frisby and M. Featherstone, London, Sage

Sinclair, I. (2003) *London Orbital*, Harmondsworth, Penguin

Skelton, T. and Allen, T., eds (1999) *Culture and Global Change*, London, Routledge

Smith, D. (1988) *The Everyday World as Problematic: A Feminist Sociology*, Milton Keynes, Open University Press

Smith, M. P. (1980) *The City and Social Theory*, Oxford, Blackwell

Smith, M. P. (2001) *Transnational Urbanism: Locating Globalization*, Oxford, Blackwell

Smith, N. (1996) *The New Urban Frontier*, London, Routledge

Smith, S. A. (2002) *The Russian Revolution: A Very Short Guide*, Oxford, Oxford University Press

Soja, E. (1989) *Postmodern Geographies*, London, Verso

Soja, E. (1996) *Thirdspace: Journeys to Los Angeles and Other Real and Imagined Places*, Oxford, Blackwell

Soja, E. (2000) *Postmetropolis: Critical Studies of Cities and Regions*, Oxford, Blackwell

Soleri, P. (1969) *Arcology – City in the Image of Man*, Cambridge (MA), MIT Press

Soper, K. (2005) 'The Awfulness of the Actual: Counter-consumerism in a New Age of War', *Radical Philosophy*, vol. 135, pp. 2–7

Soria y Puig, A., ed. (1999) *Cerda: The Five Bases of the General Theory of Urbanization*, Madrid, Electa

Sorkin, M. (1992) 'See You in Disneyland', in Sorkin, M. (ed.) *Variations on a Theme Park – The New American City and the End of Public Space*, New York, Noonday Press, pp. 205–232

Spivak, G. C., ed. (1988) *In Other Worlds: Essays in Cultural Politics*, London, Routledge

Stanley, N. (2000) 'Moving People, Moving Experiences: Novel Strategies in Museum Practice', in Cole (2000), pp. 42–48

Starhawk (2002) *Webs of Power: Notes from the Global Uprising*, Gabriola Island (BC), New Society

Steele, J. (1997) *An Architecture for the People: The Complete Works of Hassan Fathy*, London, Thames and Hudson

Steiner, C. B. (1994) *African Art in Transit*, Cambridge, Cambridge University Press

Stefan, P. (1974) *Paul Verlaine and the Decadence 1882–90*, Manchester, Manchester University Press

Stevens, M. A., ed. (1993) *The Impressionist and the City: Pissarro's Series Paintings*, London, Royal Academy of Arts and New Haven (CT), Yale University Press

Stitziel, J. (2005) *Fashioning Socialism: Clothing, Politics and Consumer Culture in East Germany*, Oxford, Berg

Strelow, H., ed. (1999) *Natural Reality* [exhibition catalogue], Stuttgart, Daco-Verlag

Sutcliffe, A. (1981) *Towards the Planned City: Germany, Britain, the United States and France, 1780–1914*, Oxford, Oxford University Press

Szerszynski, B. and Urry, J. (2002) 'Cultures of Consumption', *Sociological Review*, vol. 50, pp. 461–481

Tadini, E. (1997) 'Mouston, Disneyworld', *Rassegna*, vol. 70 (November), pp. 58–59

Tafuri, M. (1976) *Architecture and Utopia*, Cambridge (MA), MIT Press

Tajbakhsh, K. (2001) *The Promise of the City: Space, Identity, and Politics in Contemporary Social Thought*, Berkeley (CA), University of California Press

Tan, H. H. (2001) *Mammon Inc.*, London, Michael Joseph

Tan, S. and Yeoh, B. S. A. (2006) 'Negotiating Cosmopolitanism in Singapore's Fictional Landscape', in Binnie et al. (2006), pp. 146–167

Tang, X. (2000) *Chinese Modern: The Heroic and the Quotidian*, Durham (NC), Duke University Press

Tate Gallery (1970) *Léger and Purist Paris* [exhibition catalogue], London, Tate Gallery

Taylor, B. (1994) 'From Penitentiary to "Temple of Art": Early Metaphors of Improvement at the Millbank Tate', in Pointon (1994), pp. 9–32

Teedon, P. (2002) 'New Urban Spaces: Regenerating a Design Ethos', in Rugg and Hinchcliffe (2002), pp. 49–59

Thrasher, F. (1927) *The Gang*, Chicago, Chicago University Press

Tomlinson, J. (1999) 'Globalised Culture: The Triumph of the West?', in Skelton and Allen (1999), pp. 22–29

Tönnies, F. (1895) 'Considérations sur l'histoire moderne', *Annales de l'Institut International de Sociologie*, vol. 1, pp. 245–252

Tönnies, F. (1940) *Fundamental Concepts of Sociology*, trans. C. P. Loomis, New York, American Book Company

Tönnies, F. (1955) *Community and Association*, trans. C. P. Loomis, London, Routledge

Toon, I. (2000) 'Finding a Space in the Street: CCTV Surveillance and Young People's Use of Urban Public Space', in Bell and Haddour (2000), pp. 141–165

Tormey, S. (2005) 'From Utopian Worlds to Utopian Spaces: Reflections on the Contemporary Radical Imaginary and the Social Forum Process', *Ephemera*, vol. 5, 2, pp. 394–408

Toulmin, S. (1990) *Cosmopolis, the Hidden Agenda of Modernity*, Chicago, University of Chicago Press

Tournier, M. (1997) [1967] *Friday*, Baltimore (MD), Johns Hopkins University Press

Trowell, J. (2000) 'The Snowflake in Hell and the Baked Alaska: Improbability, Intimacy and Change in the Public Realm', in Bennett and Butler (2000), pp. 99–109

Tully, J. (1995) *Strange Multiplicity: Constitutionalism in an Age of Diversity*, Cambridge, Cambridge University Press

Umenyilora, C. (2000) 'Empowering the Self-Builder', in Hughes and Sadler (2000), pp. 210–221

UNESCO (1996) *Our Creative Diversity: Report of the World Commission on Culture and Development* [summary version], Paris, UNESCO Culture and Development Co-ordination Office

Universal Forum of Cultures (nd) *Universal Forum of Cultures Barcelona 2004*, Barcelona, Universal Forum of Cultures

Universal Forum of Cultures (2003) *Forum Barcelona 2004*, Barcelona, Universal Forum of Cultures

Urry, J. (1990) *The Tourist Gaze: Leisure and Travel in Contemporary Societies*, London, Sage

Urry, J. (1995) *Consuming Places*, London, Routledge

Vale, B. and Vale, R. (1991) *Green Architecture: Design for a Sustainable Future*, London, Thames and Hudson

Veblen, T. (1970) [1899] *Theory of the Leisure Class*, London, Unwin

Vergo, P. (1975) *Art in Vienna 1898–1918*, London, Phaidon

Wackernagel, M. and Rees, W. (1996) *Our Ecological Footprint: Reducing Human Impact on the Earth*, Gabriola Island (BC), New Society Publishers

Waelti-Walters, J. (1996) *Jeanne Hyvrard: Theorist of the Modern World*, Edinburgh, Edinburgh University Press

Wall, D. (1999) *Earth First! and the Anti-Roads Movement: Radical Environmentalism and Comparative Social Movements*, London, Routledge

Warde, A. (2002) 'Production, Consumption and the Cultural Economy', in du Gay and Pryke (2002), pp. 185–200

Warner, M. (1985) *Monuments and Maidens: The Allegory of the Female Form*, London, Picador

Watson, G. (1997) *Against the Mega-Machine: Essays on Empire and its Enemies*, New York, Autonomedia

Welsch, W. (1997) *Undoing Aesthetics*, London, Sage

Westwood, S. and Williams, J., eds (1997) *Imagining Cities: Scripts, Signs, Memory*, London, Routledge

Wigglesworth, S. and Till, J., eds (1998) 'The Everyday and Architecture', *Architectural Design*, profile 134, July/August

Wigley, M. (1992) 'Untitled: The Housing of Gender', in Colomina (1992), pp. 327–389

Willett, J. (1967) *Art in a City*, London, Methuen

Willett, J. (1978) *The New Sobriety: Art and Politics in the Weimar Period 1917–33*, London, Thames and Hudson

Williams, N., Herrmann, D. and Kemp, C. (2003) *Estonia, Latvia & Lithuania* [Lonely Planet Guide], Footscray (Victoria), Lonely Planet Publications

Williams, R. (1965) *The Long Revolution*, Harmondsworth, Penguin [first published, London, Chatto & Windus, 1961]

Williams, R. (1976) *Keywords: A Vocabulary of Culture and Society*, London, Fontana

Williams, R. (1980) *Problems in Materialism and Culture*, London, Verso

Williams, R. (1983) *Culture and Society: 1780–1950*, New York, Columbia University Press [first published, 1958]

Williams, R. (1988) *Marxism and Literature*, Oxford, Oxford University Press

Williams, R. (1989) *The Politics of Modernism*, ed. Tony Pinkney, London, Verso

Wilson, A. (1988) 'Eritrea: The Experience of Creating a New Culture', in Owusu (1988), pp. 195–204

Wilson, E. (1991) *The Sphinx in the City: Urban Life, the Control of Disorder, and Women*, Berkeley (CA), University of California Press

Wilson, E. (1992) 'The Invisible Flâneur', *New Left Review*, vol. 191, pp. 90–100

Wilson, E. (2000) *Bohemians: The Glamorous Outcasts*, London, I. B. Tauris

Wilson, E. (2001) *The Contradictions of Culture: Cities, Culture, Women*, London, Sage

Wilson, W. H. (1989) *The City Beautiful Movement*, Baltimore (MD), Johns Hopkins University Press

Wirth, L. (1964) [1948] *On Cities and Social Life*, Chicago, University of Chicago Press

Wirth, L. (2003) 'Urbanism as a Way of Life', in LeGates and Stout (2003), pp. 98–105 [first published in *American Journal of Sociology*, XLIV, 1 (July 1938)]

Witkin, R. W. (2003) *Adorno on Popular Culture*, London, Routledge

Wolff, J. (1991) 'The Invisible *Flâneuse*: Women and the Literature of Modernity', in Benjamin (1991), pp. 141–156

Wolin, R. (1994) *Walter Benjamin: An Aesthetic of Redemption*, Berkeley (CA), University of California Press

Wood, A. (2002) 'Re-Reading Disney's Celebration: Gendered Topography in a Heterotopian Pleasure Garden', in Bingaman, Sanders and Zorach (2002), pp. 188–203

Wood, P., ed. (1999) *The Challenge of the Avant-Garde*, New Haven (CT), Yale University Press

Worsley, P. (1999a) 'Classic Concepts of Culture', in Skelton and Allen (1999), pp. 13–21

Worsley, P. (1999b) 'Culture and Development Theory', in Skelton and Allen (1999), pp. 30–41

Yeoh, B. S. A. (2005) 'The Global Cultural City? Spatial Imagineering and Politics in the (Multi)cultural Marketplaces of South-east Asia', *Urban Studies*, vol. 42, 5/6, pp. 945–958

Young, I. M. (1990) *Justice and the Politics of Difference*, Princeton (NJ), Princeton University Press

Young, I. M. (2000) *Inclusion and Democracy*, Oxford, Oxford University Press

Younge, G. (1988) *Art of the South African Townships*, London, Thames and Hudson

Yúdice, G. (2003) *The Expediency of Culture: Uses of Culture in the Global Era*, Durham (NC), Duke University Press

Zucker, P. (1959) *Town and Square*, Cambridge (MA), MIT Press

Zukin, S. (1989) *Loft-Living: Cultural Capital in Urban Change*, New Brunswick (NJ), Rutgers University Press

Zukin, S. (1991) *Landscapes of Power: From Detroit to Disney World*, Berkeley (CA), University of California Press

Zukin, S. (1995) *The Cultures of Cities*, Oxford, Blackwell

Zukin, S. (1996a) 'Space and Symbols in an Age of Decline', in King (1996), pp. 43–59

Zukin, S. (1996b) 'Cultural Strategies of Economic Development and the Hegemony of Vision', in Merrifield and Swyngedouw (1996), pp. 223–243

Index